Nadia Nedjah, Leandro dos Santos Coelho, and Luiza de Macedo de Mourelle (Eds.)

Multi-Objective Swarm Intelligent Systems

Studies in Computational Intelligence, Volume 261

Editor-in-Chief
Prof. Janusz Kacprzyk
Systems Research Institute
Polish Academy of Sciences
ul. Newelska 6
01-447 Warsaw
Poland
E-mail: kacprzyk@ibspan.waw.pl

Further volumes of this series can be found on our homepage: springer.com

Vol. 241. János Fodor and Janusz Kacprzyk (Eds.)
Aspects of Soft Computing, Intelligent Robotics and Control, 2009
ISBN 978-3-642-03632-3

Vol. 242. Carlos Artemio Coello Coello, Satchidananda Dehuri, and Susmita Ghosh (Eds.)
Swarm Intelligence for Multi-objective Problems in Data Mining, 2009
ISBN 978-3-642-03624-8

Vol. 243. Imre J. Rudas, János Fodor, and Janusz Kacprzyk (Eds.)
Towards Intelligent Engineering and Information Technology, 2009
ISBN 978-3-642-03736-8

Vol. 244. Ngoc Thanh Nguyen, Rados law Piotr Katarzyniak, and Adam Janiak (Eds.)
New Challenges in Computational Collective Intelligence, 2009
ISBN 978-3-642-03957-7

Vol. 245. Oleg Okun and Giorgio Valentini (Eds.)
Applications of Supervised and Unsupervised Ensemble Methods, 2009
ISBN 978-3-642-03998-0

Vol. 246. Thanasis Daradoumis, Santi Caballé, Joan Manuel Marquès, and Fatos Xhafa (Eds.)
Intelligent Collaborative e-Learning Systems and Applications, 2009
ISBN 978-3-642-04000-9

Vol. 247. Monica Bianchini, Marco Maggini, Franco Scarselli, and Lakhmi C. Jain (Eds.)
Innovations in Neural Information Paradigms and Applications, 2009
ISBN 978-3-642-04002-3

Vol. 248. Chee Peng Lim, Lakhmi C. Jain, and Satchidananda Dehuri (Eds.)
Innovations in Swarm Intelligence, 2009
ISBN 978-3-642-04224-9

Vol. 249. Wesam Ashour Barbakh, Ying Wu, and Colin Fyfe
Non-Standard Parameter Adaptation for Exploratory Data Analysis, 2009
ISBN 978-3-642-04004-7

Vol. 250. Raymond Chiong and Sandeep Dhakal (Eds.)
Natural Intelligence for Scheduling, Planning and Packing Problems, 2009
ISBN 978-3-642-04038-2

Vol. 251. Zbigniew W. Ras and William Ribarsky (Eds.)
Advances in Information and Intelligent Systems, 2009
ISBN 978-3-642-04140-2

Vol. 252. Ngoc Thanh Nguyen and Edward Szczerbicki (Eds.)
Intelligent Systems for Knowledge Management, 2009
ISBN 978-3-642-04169-3

Vol. 253. Roger Lee and Naohiro Ishii (Eds.)
Software Engineering Research, Management and Applications 2009, 2009
ISBN 978-3-642-05440-2

Vol. 254. Kyandoghere Kyamakya, Wolfgang A. Halang, Herwig Unger, Jean Chamberlain Chedjou, Nikolai F. Rulkov, and Zhong Li (Eds.)
Recent Advances in Nonlinear Dynamics and Synchronization, 2009
ISBN 978-3-642-04226-3

Vol. 255. Catarina Silva and Bernardete Ribeiro
Inductive Inference for Large Scale Text Classification, 2009
ISBN 978-3-642-04532-5

Vol. 256. Patricia Melin, Janusz Kacprzyk, and Witold Pedrycz (Eds.)
Bio-inspired Hybrid Intelligent Systems for Image Analysis and Pattern Recognition, 2009
ISBN 978-3-642-04515-8

Vol. 257. Oscar Castillo, Witold Pedrycz, and Janusz Kacprzyk (Eds.)
Evolutionary Design of Intelligent Systems in Modeling, Simulation and Control, 2009
ISBN 978-3-642-04513-4

Vol. 258. Leonardo Franco, David A. Elizondo, and José M. Jerez (Eds.)
Constructive Neural Networks, 2009
ISBN 978-3-642-04511-0

Vol. 259. Kasthurirangan Gopalakrishnan, Halil Ceylan, and Nii O. Attoh-Okine (Eds.)
Intelligent and Soft Computing in Infrastructure Systems Engineering, 2009
ISBN 978-3-642-04585-1

Vol. 260. Edward Szczerbicki and Ngoc Thanh Nguyen (Eds.)
Smart Information and Knowledge Management, 2009
ISBN 978-3-642-04583-7

Vol. 261. Nadia Nedjah, Leandro dos Santos Coelho, and Luiza de Macedo de Mourelle (Eds.)
Multi-Objective Swarm Intelligent Systems, 2009
ISBN 978-3-642-05164-7

Nadia Nedjah, Leandro dos Santos Coelho, and
Luiza de Macedo de Mourelle (Eds.)

Multi-Objective Swarm Intelligent Systems

Theory & Experiences

Springer

Nadia Nedjah
Universidade do Estado do Rio de Janeiro
Faculdade de Engenharia
sala 5022-D
Rua São Francisco Xavier 524
20550-900, MARACANÃ-RJ
Brazil
E-mail: nadia@eng.uerj.br

Luiza de Macedo Mourelle
Universidade do Estado do Rio de Janeiro
Faculdade de Engenharia
sala 5022-D
Rua São Francisco Xavier 524
20550-900, MARACANÃ-RJ
Brazil
E-mail: ldmm@eng.uerj.br

Leandro dos Santos Coelho
Universidade Federal do Paraná
Departamento de Engenharia Elétrica
Pós-Graduação em Engenharia Elétrica
Centro Politécnico
81531-980, Curitiba-PR, Brazil
and
Pontifícia Universidade
Católica do Paraná
Centro de Ciências
Exatas de Tecnologia
Pós-Graduação em Engenharia
de Produção e Sistemas (PPGEPS)
Rua Imaculada Conceição 1155
80215-901, Curitiba-PR, Brazil
E-mail: leandro.coelho@pucpr.br

ISBN 978-3-642-26151-0
e-ISBN 978-3-642-05165-4
DOI 10.1007/978-3-642-05165-4
Studies in Computational Intelligence
ISSN 1860-949X

Softcover reprint of the hardcover 1st edition 2010

Typeset & Cover Design: Scientific Publishing Services Pvt. Ltd., Chennai, India.

Printed in acid-free paper

9 8 7 6 5 4 3 2 1

springer.com

Preface

Recently, a new class of heuristic techniques, the swarm intelligence has emerged. In this context, more recently, biologists and computer scientists in the field of "artificial life" have been turning to insects for ideas that can be used for heuristics. Many aspects of the collective activities of social insects, such as foraging of ants, birds flocking and fish schooling are self-organizing, meaning that complex group behavior emerges from the interactions of individuals who exhibit simple behaviors by themselves.

Swarm intelligence is an innovative computational way to solving hard problems. This discipline is mostly inspired by the behavior of ant colonies, bird flocks and fish schools and other biological creatures. In general, this is done by mimicking the behavior of these swarms.

Swarm intelligence is an emerging research area with similar population and evolution characteristics to those of genetic algorithms. However, it differentiates in emphasizing the cooperative behavior among group members. Swarm intelligence is used to solve optimization and cooperative problems among intelligent agents, mainly in artificial network training, cooperative and/or decentralized control, operational research, power systems, electro-magnetics device design, mobile robotics, and others. The most well-known representatives of swarm intelligence in optimization problems are: the food-searching behavior of ants, particle swarm optimization, and bacterial colonies.

Real-world engineering problems often require concurrent optimization of several design objectives, which are conflicting in most of the cases. Such an optimization is generally called multi-objective or multi-criterion optimization. In this context, the development of improvements for swarm intelligence methods to multi-objective problems is an emergent research area.

In Chapter 1, which is entitled *Multi-objective Gaussian Particle Swarm Approach Applied to Multi-Loop PI Controller Tuning of a Quadruple-Tank System*, the authors propose a multi-objective particle swarm optimization approach inspired from some previous related work. The approach updates the velocity vector using the Gaussian distribution, called MGPSO, to solve the multi-objective optimization of the multi-loop Proportional-Integral control tuning.

In Chapter 2, which is entitled *A Non-Ordered Rule Induction Algorithm Through Multi-Objective Particle Swarm Optimization: Issues and Applications*, the authors propose a new approach, called MOPSO-N, validate its efficiency. They also describe the application of MOPSO-N in the Software Engineering domain.

In Chapter 3, which is entitled *Use of Multiobjective Evolutionary Algorithms in Water Resources Engineering*, the authors investigate the efficiency of multi-objective particle swarm optimizaton in Water Resources Engineering.

In Chapter 4, which is entitled *Micro-MOPSO: A Multi-Objective Particle Swarm Optimizer that Uses a Very Small Population Size*, the author present a multi-objective evolutionary algorithm (MOEA) based on the heuristic called "particle swarm optimization" (PSO). This multi-objective particle swarm optimizer (MOPSO) is characterized for using a very small population size, which allows it to require a very low number of objective function evaluations (only 3000 per run) to produce reasonably good approximations of the Pareto front of problems of moderate dimensionality.

In Chapter 5, which is entitled *Dynamic Multi-objective Optimisation using PSO*, the author introduce the usage of the vector evaluated particle swarm optimiser (VEPSO) to solve DMOOPs, wherein every objective is solved by one swarm and the swarms share knowledge amongst each other about the objective that it is solving.

In Chapter 6, which is entitled *Meta-PSO for Multi-Objective EM Problems*, the authors investigate some variations over the standard PSO algorithm, referred to as Meta-PSO, aiming at enhancing the global search capability, and, therefore, improving the algorithm convergence.

In Chapter 7, which is entitled *Multi-Objective Wavelet-Based Pixel-Level Image Fusion Using Multi-Objective Constriction Particle Swarm Optimization*, the authors present a new methodology of multi-objective pixel-level image fusion based on discrete wavelet transform and design an algorithm of multi-objective constriction particle swarm optimization (MOCPSO).

In Chapter 8, which is entitled *Multi-objective Damage Identification Using Particle Swarm Optimization Techniques*, the authors present a particle swarm optimization-based strategies for multi-objective structural damage identification. Different variations of the conventional PSO based on evolutionary concepts are implemented for detecting the damage of a structure in a multi-objective framework.

The editors are very much grateful to the authors of this volume and to the reviewers for their tremendous service by critically reviewing the chapters. The editors would like also to thank Prof. Janusz Kacprzyk, the editor-in-chief of the Studies in Computational Intelligence Book Series and Dr. Thomas Ditzinger, Springer Verlag, Germany for the editorial assistance and excellent cooperative collaboration to produce this important scientific work.

We hope that the reader will share our excitement to present this volume on **Multi-Objective Swarm Intelligent Systems** and will find it useful.

August 2009

Nadia Nedjah
State University of Rio de Janeiro, Brazil

Leandro dos S. Coelho
Federal University of Paraná, Brazil and
Pontifical Catholic University of Paraná, Brazil

Luiza M. Mourelle
State University of Rio de Janeiro, Brazil

Contents

List of Figures

List of Tables

1

Multiobjective Gaussian Particle Swarm Approach Applied to Multi-loop PI Controller Tuning of a Quadruple-Tank System

Leandro dos Santos Coelho[1], Helon Vicente Hultmann Ayala[2], Nadia Nedjah[3], and Luiza de Macedo Mourelle[4]

[1] Federal University of Paraná, Department of Electrical Engineering
Zip code 81531-980, Curitiba, Paraná, Brazil, and
Industrial and Systems Engineering Graduate Program
Pontifical Catholic University of Paraná, PPGEPS/PUCPR
Imaculada Conceição, 1155, Zip code 80215-901, Curitiba, Paraná, Brazil
leandro.coelho@pucpr.br

[2] Undergraduate Program at Mechatronics Engineering
Pontifical Catholic University of Paraná
Imaculada Conceição, 1155, Zip code 80215-901, Curitiba, Paraná, Brazil
helonayala@gmail.com

[3] Department of Electronics Engineering and Telecommunications,
Engineering Faculty,
State University of Rio de Janeiro,
Rua São Francisco Xavier, 524, Sala 5022-D,
Maracanã, Rio de Janeiro, Brazil
nadia@eng.uerj.br
http://www.eng.uerj.br/~nadia

[4] Department of System Engineering and Computation,
Engineering Faculty,
State University of Rio de Janeiro,
Rua São Francisco Xavier, 524, Sala 5022-D,
Maracanã, Rio de Janeiro, Brazil
ldmm@eng.uerj.br
http://www.eng.uerj.br/~ldmm

The use of PI (Proportional-Integral), PD (Proportional-Derivative) and PID (Proportional-Integral-Derivative) controllers have a long history in control engineering and are acceptable for most of real applications because of their simplicity in architecture and their performances are quite robust for a wide range of operating conditions. Unfortunately, it has been quite difficult to tune properly the gains of PI, PD, and PID controllers because many industrial plants are often burdened with problems such as high order, time delays, and nonlinearities. Recently, several metaheuristics, such as evolutionary algorithms, swarm intelligence and simulated annealing, have been proposed for the tuning of mentioned controllers. In this context, different metaheuristics have recently received much interest for achieving high efficiency and searching global

N. Nedjah et al. (Eds.): Multi-Objective Swarm Intelligent Systems, SCI 261, pp. 1–16.
springerlink.com

optimal solution in problem space. Multi-objective evolutionary and swarm intelligence approaches often find effectively a set of diverse and mutually competitive solutions. A multi-loop PI control scheme based on a multi-objective particle swarm optimization approach with updating of velocity vector using Gaussian distribution (MGPSO) for multi-variable systems is proposed in this chapter. Particle swarm optimization is a simple and efficient population-based optimization method motivated by social behavior of organisms such as fish schooling and bird flocking. The proposal of PSO algorithm was put forward by several scientists who developed computational simulations of the movement of organisms such as flocks of birds and schools of fish. Such simulations were heavily based on manipulating the distances between individuals, i.e., the synchrony of the behavior of the swarm was seen as an effort to keep an optimal distance between them. In theory, at least, individuals of a swarm may benefit from the prior discoveries and experiences of all the members of a swarm when foraging. The fundamental point of developing PSO is a hypothesis in which the exchange of information among creatures of the same species offers some sort of evolutionary advantage. PSO demonstrates good performance in many function optimization problems and parameter optimization problems in recent years. Application of the proposed MGPSO using concepts of Pareto optimality to a multi-variable quadruple-tank process is investigated in this paper. Compared to a classical multi-objective PSO algorithm which is applied to the same process, the MGPSO shows considerable robustness and efficiency in PI control tuning.

1.1 Introduction

In many fields of science, the procedure of optimization sometimes has more than one objective, thus the need for multi-objective optimization is obvious. One of the factors that differentiate single objective optimization when compared to multi-objective optimization is that the optimum solution for multi-objective optimization is not necessarily unique. In general, the multi-objective variant of a problem is harder than the single objective case.

In a typical multi-objective optimization problem (also known as multi-criterion optimization), there is a family of equivalent solutions that are superior to the rest of the solutions and are considered equal from the perspective of simultaneous optimization of multiple (and possibly competing) objective functions. In other words, in multi-objective optimization there is no single optimal solution. Instead, the interaction of multiple objectives yields a set of efficient (non inferior) or non-dominated solutions, known as Pareto-optimal solutions, which give a decision maker more flexibility in the selection of a suitable alternative.

In other words, the multi-objective optimizer is expected to give a set of all representative equivalent and diverse solution. Objectives to be simultaneously optimized may be mutually conflicting. Additionally, achieving proper diversity in the solutions while approaching convergence is another challenge in multi-objective optimization. Different evolutionary algorithms and swarm intelligence approaches have been validated for multi-objective optimization problems. Evolutionary algorithms and swarm intelligence approaches usually do not guarantee to identify optimal trade-offs but try to find a good approximation, i.e., a set of solutions whose objective vectors are (hopefully) not too far away from the optimal objective vectors [38]. Various multi-objective evolutionary [36, 9, 7, 8] and swarm intelligence [34, 24, 1] approaches are available, and certainly we are interested in the technique that provides the best approximation for a given problem.

Many metaheuristics based on evolutionary and swarm intelligence paradigms have been used for multi-objective optimization problems in process engineering [4, 25, 23, 37, 2, 30]. This is mainly due to their ability to *(i)* find multiple solutions in a given run; *(ii)* work without derivatives information; and *(iii)* efficiently converge to a potential solution.

In recent years, there has been an increased interest in the study, design and analysis of particle swarm optimization (PSO) approaches to solve the multiobjective optimization problems. Due to its fast convergence, PSO has been advocated to be especially suitable for multiobjective optimization. PSO is a population-based approach of swarm intelligence field that was first developed by James Kennedy and Russell Eberhart [18, 12]. Their original idea was to simulate the social behavior of a flock of birds trying to reach an unknown destination (fitness function), e.g., the location of food resources when flying through the field (search space). In other words, PSO is inspired by adaptation of a natural system based on the metaphor of social communication and interaction. Despite its simplicity, PSO provides efficient yet accurate solutions to many multiobjective engineering problems [34, 24, 1].

In this chapter, we propose a multiobjective particle swarm optimization approach inspired from [31], however, with updating of velocity vector using Gaussian distribution (MGPSO) to solve the multiobjective optimization of the multi-loop PI (Proportional-Integral) control tuning. Furthermore, simulation results of multi-loop PI control using MGPSO to a multivariable quadruple-tank process are presented and discussed.

The remaining portion of the paper are organized as follows: in Section 1.2, a description of quadruple-tank process is provided. Section 1.3 presented the fundamentals of PSO and MGPSO approaches. Simulation results are drawn in Section 1.4. Last but not least, we draw some conclusions in Section 1.5.

1.2 Description of Quadruple-Tank Process

Several researchers have investigated the problem of controlling liquid flow of a single or multiple tanks [14, 29, 20]. The quadruple-tank introduced in [17] has received a great attention because it presents interesting properties in both control education and research. The quadruple-tank exhibits in an elegant and simple way complex dynamics. Such dynamic characteristics include interactions and a transmission zero location that are tunable in operation [11].

The quadruple-tank process consists of four interconnected water tanks and two pumps. Its inputs are v_1 and v_2 (input voltages to the pumps) and the outputs are y_1 and y_2 (voltages from level measurement devices) [17]. The quadruple-tank process can easily be build by using two double-tank processes, which are standard processes in many control laboratories. The schematic diagram of quadruple-tank process is presented in Figure 1.1.

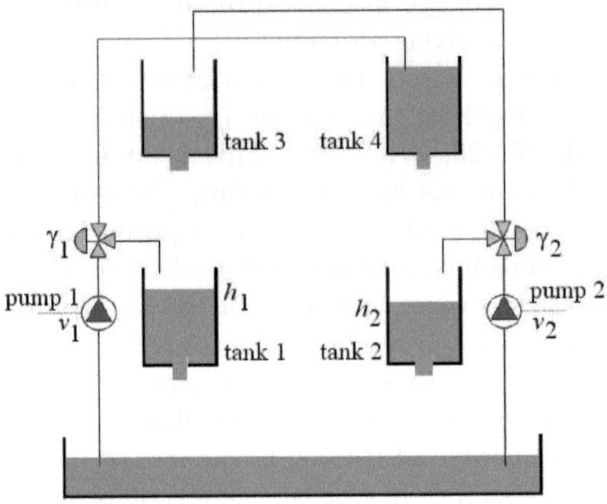

Fig. 1.1. Schematic diagram of quadruple-tank process

For this process, mass balances and Bernoullis law yield [17] (1.1)–(1.4).

$$\frac{dh_1}{dt} = -\frac{a_1}{A_1}\sqrt{2gh_1} + \frac{a_3}{A_1}\sqrt{2gh_3} + \frac{\gamma_1 k_1}{A_1}v_1, \quad (1.1)$$

$$\frac{dh_2}{dt} = -\frac{a_2}{A_2}\sqrt{2gh_2} + \frac{a_4}{A_2}\sqrt{2gh_4} + \frac{\gamma_2 k_2}{A_2}v_2, \quad (1.2)$$

$$\frac{dh_3}{dt} = -\frac{a_3}{A_3}\sqrt{2gh_3} + \frac{(1-\gamma_2)k_2}{A_3}v_2, \quad (1.3)$$

$$\frac{dh_4}{dt} = -\frac{a_4}{A_4}\sqrt{2gh_4} + \frac{(1-\gamma_1)k_2}{A_4}v_1, \quad (1.4)$$

where A_i is the cross-section of tank i, a_i is the cross-section of outlet hole and h_i is the water level. The voltage applied to pump i is v_i, and the corresponding flow is $k_i v_i$. The parameters $\gamma_1, \gamma_2 \in (0, 1)$ are determined from how the valves are set. The flow to Tank 1 is $\gamma_1 k_1 v_1$ and the flow to tank 4 is $(1-\gamma_1)k_1 v_1$ and similarly to tank 2 and tank 3. The acceleration of gravity is denoted as g. The measured level signals are $k_c h_1$ and $k_c h_2$. These signal represented the outputs signals y, i.e. $y_1(t) = k_c h_1(t)$ and $y_2(t) = k_c h_2(t)$, where t represents the time. The adopted time sampling in this work was 1 s. The parameter values used in this paper, as in [17], are given in Table 1.1.

Table 1.1. Parameters values adopted for the quadruple-tank process

Parameter	Unit	Value
A_1, A_3	cm^2	28
A_2, A_4	cm^2	32
a_1, a_3	cm^2	0.071
a_1, a_4	cm^2	0.057
k_c	V/cm	0.50
g	cm/s^2	981

The PID controller is a standard tool for industrial automation. The flexibility, simple structure and robustness of these controllers make it possible to use PID control in many applications. However, many control problems can be handled very well by PI control without derivative action of PID. In terms of two multi-loop PI controllers, in this work is considered $K(s)$ with the classical structure of (1.5).

$$K(s) = \begin{bmatrix} k_{11}(s) & \ldots & k_{1n}(s) \\ \vdots & \ddots & \vdots \\ k_{n1}(s) & \ldots & k_{nn}(s) \end{bmatrix} \tag{1.5}$$

where $n = 2$. A continuous-time multi-loop PI controller is usually given in the form of $k_{ij}(s)$, $i, j \in n = \{1, 2, \ldots, n\}$, as defined in (1.6).

$$k_{ij}(s) = K_{p_{ij}} \left(1 + \frac{1}{T_{i_{ij}}} \right), \tag{1.6}$$

wherein $K_{(p_{ij})}$ is the proportional gain, and $T_{i_{ij}}$ is the integral gain.

The key of designing a PI controller is to determine two PI control gains. In this work, a multi-loop PI control was adopted, where the proportional and integral gains are null for $i \neq j$. In this context, a discretization of equation (1.6) using forward rectangular method (details in chapter 4 of [3]) to design a digital multi-loop PI control with incremental form is realized.

The MOPSO proposed in [31] and the MGPSO must search the parameters of a 2×2 decoupled PI, i.e., search the parameters $K_{p_{11}}$, $K_{p_{22}}$, $T_{i_{11}}$ and $T_{i_{22}}$.

1.3 Fundamentals of Multi-objective Optimization and PSO

This section presents the fundamentals of multi-objective optimization and PSO. First, a brief overview of the multi-objective optimization is provided, and finally the design of the MOPSO proposed by [31] and the MGPSO algorithm are discussed.

1.3.1 Multi-objective Optimization

In contrast to single-objective optimization, it is essential to obtain a well-distributed and diverse solution set for finding the final tradeoff in multi-objective optimization. Multiobjective optimization can be defined as the problem of finding a vector of decision variables that satisfies constraints and optimizes a vector function whose elements represent the objective functions. A general multiobjective optimization problem containing a number of objectives to be minimized and (optional) constraints to be satisfied can be written as in (1.7).

$$\begin{array}{ll} \text{Minimize} & f_m(X),\ m = 1, 2, \ldots, M \\ \text{subject to constraint } g_k(X) \leq c_k, & k = 1, 2, \ldots, K \end{array} \tag{1.7}$$

where $X = \{x_n, n = 1, 2, \ldots, N\}$ is a vector of decision variables and $F = \{f_m, m = 1, 2, \ldots, M\}$ are M objectives to be minimized [26].

In a typical multi-objective optimization problem, there exists a family of equivalent solutions that are superior to the rest of the solutions and are considered equal from the perspective of simultaneous optimization of multiple (and possibly competing) objective functions. Such solutions are called non inferior, non dominated, or Pareto-optimal solutions, and are such that no objective can be improved without degrading at least one of the others, and, given the constraints of the model, no solution exist beyond the true Pareto front. The goal of multi-objective algorithms is to locate the (whole) Pareto front. Clearly, the Pareto front is the image of the Pareto optimal set in the objective space.

Each objective component of any non dominated solution in the Pareto optimal set can only be improved by degrading at least one of its other objective components. A vector fa is said to dominate another vector f_b, denoted as in (1.8).

$$f_a \prec f_b \text{ iff } f_{a,i} \leq f_{b,i}, \forall i = \{1, 2, \ldots, M\}, \text{ and }, \text{ where } f_{a,i} \prec f_{b,i} \tag{1.8}$$

Summarizing, there are two goals in multi-objective optimization: *(i)* to discover solutions as close to the Pareto-front as possible, and *(ii)* to find solutions as diverse as possible in the obtained nondominated front.

Methods of multi-objective optimization can be classified in many ways according to different criteria. In [16] the methods are classified according to the participation of the decision maker in the solution process. The classes are: *(i)* methods where no articulation of preference information is used (no-preference methods); *(ii)* methods where a posteriori articulation of preference information used (a posteriori methods); *(iii)* methods where a priori articulation of preference information used (a priori methods); and *(iv)* methods where progressive articulation of preference information is used (interactive methods). The method adopted in this paper is the posteriori method.

In posteriori methods, after the Pareto optimal set has been generated, it is presented to the decision maker, who selects the most preferred among the alternatives.

1.3.2 Classical PSO and MOPSO Approaches

PSO is in principle such a multi-agent parallel search technique. The PSO consists of three steps, namely, generating positions and velocities of particles, velocity update, and finally, position update. PSO is easy to implement in computer simulations using basic mathematical and logic operations, since its working mechanism only involves two fundamental updating rules. Particles are conceptual entities that constitute a swarm, which fly through the multi-dimensional search space. The relationship between swarm and particles in PSO is similar to the relationship between population and chromosomes in genetic algorithm. At any particular instant each particle has a position and a velocity. The position vector of a particle with respect to the origin of the search space represents a trial solution of the search problem.

These particles fly with a certain velocity and find the global best position after some iteration. At each iteration, each particle can adjust its velocity vector, based on its momentum and the influence of its best position (*pbest* – personal best) as well as the best position of its neighbors (*gbest* - global best), and then compute a new position that the "particle" is to fly to. On other words, it finds the global optimum by simply adjusting the trajectory of each individual towards its own best location and towards the best particle of the swarm at each generation of evolution. The swarm direction of a particle is defined by the set of particles neighboring the particle and its history experience.

The unpublished manuscript [27] proposed the first extension of the PSO strategy for solving multi-objective problems. There have been several recent fundamental proposals using PSO to handle multiple objectives, surveyed in [34].

However, the high speed of convergence in MOPSO approaches often implies a rapid loss of diversity during the optimization process. In this context, several MOPSO have difficulties in controlling the balance between explorations and exploitations.

In [31], the authors propose a multiobjective PSO (MOPSO) incorporating the concept of nearest neighbor density estimator for selecting the global best particle and also for deleting particles from the external archive of non-dominated solutions. When selecting a leader, the archive of nondominated solutions is sorted in descending order with respect to the density estimator, and a particle is randomly chosen from the top part of the list. On the other hand, when the external archive is full, it is again sorted in descending order with respect to the density estimator value and a particle is randomly chosen to be deleted, from the bottom part of the list. This approach uses the mutation operator proposed in [6] in such a way that it is applied only during a certain number of generations at the beginning of the process. Finally, the authors adopt the constraint-handling technique from the NSGA-II [10].

The procedure for implementing the MOPSO given in [31] is given by the following steps:

1. Initialize a population or swarm of particles with random positions and velocities in the n dimensional problem space using uniform probability distribution function. Set the generation counter, $t = 0$;
2. Evaluate the particles and store the nondominated particles in swarm in an external archive A;
3. Compute the crowding distance values of each nondominated solution in archive A. The crowding measure of a particle i reflects the distribution of other particles around i. The smaller distance is, the more the number of individuals surrounding i is. Compared with the number of particles in a grid, the crowding measure exactly describes the relative position relation among different particles.
4. Sort the nondominated solutions in A in descending crowding distance values;
5. Randomly select the global best guide for the swarm form a specified top portion (e.g. top 10%) for the sorted archive A and store its position to *gbest*.
6. Change the velocity, v_i, and position of the particle, x_i, according to equations:

$$v_i(t+1) = w.v_i(t) + c_1.ud.[p_i(t) - x_i(t)] + c_2.Ud.[p_g(t) - x_i(t)] \quad (1.9)$$

$$x_i(t+1) = x_i(t) + \Delta t.v_i(t+1) \quad (1.10)$$

   where w is the inertia weight; $i = 1, 2, \ldots, N$ indicates the number of particles of population (swarm); $t = 1, 2, \ldots t_{max}$, indicates the generations (iterations), w is a parameter called the inertial weight; $v_i = [v_{i_1}, v_{i_2}, \ldots v_{i_n}]^T$ stands for the velocity of the ith particle, stands for the position of the ith particle of population, and represents the best previous position of the ith particle.

   Positive constants c_1 and c_2 are the cognitive and social factors, respectively, which are the acceleration constants responsible for varying the particle velocity towards *pbest* and *gbest*, respectively. Index g represents the index of the best particle among all the particles in the swarm. Variables ud and Ud are two random functions with uniform distribution in the range [0,1]. Equation 1.10 represents the position update, according to its previous position and its velocity, considering $\Delta = 1$.
7. Perform the mutation operation proposed in [6] with probability of 0.5;
8. Evaluate the particles in swarm;
9. Insert all new nondominated solution in swarm into A if they are not dominated by any of the stored solutions. All dominated solutions in the archive by the new solution are removed from the archive. If the archive is full, the solution to be replaced is determined by the following steps:

(a) compute the crowding distance values of each nondominated solution in the archive A; *(b)* sort the nondominated solutions in A in descending crowding distance values, and *(c)* randomly select a particle from a specified bottom portion (e.g. lower 10%) which comprise the most crowded particles in the archive then replace it with the new solution;
10. Increment the generation counter, $t = t + 1$;
11. Return to Step 3 until a stop criterion is met, usually a sufficiently good fitness or a maximum number of iterations, t_{max}. In this work, the t_{max} value is adopted.

1.3.3 The Proposed MGPSO Approach

Recently, several investigations have been undertaken to improve the performance of standard PSO [35, 13, 28] and MOPSO [34]. Most PSO algorithms use uniform probability distribution to generate random numbers [19]. However, recent design approaches using Gaussian probability distributions to generate random numbers to updating the velocity equation of PSO have been proposed [21, 33, 15, 22]. In this chapter, following the same line of study, it presents a new approach called MGPSO using Gaussian probability distribution.

Generating random numbers using Gaussian distribution sequences with zero mean and unit variance for the stochastic coefficients of PSO may provide a good compromise between the probability of having a large number of small amplitudes around the current points (fine tuning) and a small probability of having higher amplitudes, which may allow particles to move away from the current point and escape from local minima.

The proposed MGPSO approach uses an operator of velocity updating based on truncated Gaussian distribution [5]. In this case, Equation 1.9 is modified as in (1.11).

$$v_i(t+1) = w.v_i(t) + c_1.ud.[p_i(t) - x_i(t)] + c_2.Ud.[p_g(t) - x_i(t)] \quad (1.11)$$

where Gd are numbers generated with Gaussian distribution truncated in range [0,1].

1.4 Simulation Results

The experiments were conducted for 30 independent runs in MATLAB environment to evaluate the performance of MOPSO [31] and MGPSO on the tuning of two PI controllers in multi-loop configuration applied to the quadruple-tank process. The adopted setup for the MOPSO and MGPSO was $c_1 = c_2 = 1.0$, and the range of the inertia weight w is from 0.5 to 0.3 during the generations for the MOPSO and MGPSO approaches. The population size was 20 particles, stopping criterion, t_max, of 200 generations, and external archive size equal to 500. The search space was $K_{p_1}, K_{p_2} \in [-50, 50]$

and T_{i_1}, $T_{i_2} \in [0, 400]$. The total of samples to evaluate the fitness function is 3200 and time sampling adopted is 1 s. Unstable solutions are penalized with value of fitness equal to infinite. The adopted initial conditions are $h_1 = 12.6$, $h_2 = 13.0$, $h_3 = 4.8$, and $h_4 = 4.9$. The Runge-Kutta was the integration method employed in equations (1)–(4) during the simulations. The fitness function (minimization problem) is given in (1.12)–(1.14).

$$f = f_1 + f_2 \tag{1.12}$$

$$f_1 = \sum_{t=1}^{N} [y_{r,1}(t) - y_1(t)]^2 \tag{1.13}$$

$$f_2 = \sum_{t=1}^{N} [y_{r,2}(t) - y_2(t)]^2 \tag{1.14}$$

where $y_{r,1}$ and $y_{r,2}$ are the set-points for the output 1 and 2, y_1 and y_2 are the outputs of process.

Simulation results, presented in Figure 1.2 and Figure 1.3, showed that the non-dominated solutions of 30 runs (all solutions serve to set up the Pareto front) obtained by MOPSO with 402 solutions and MGPSO with 410 solutions. A relevant information is about the mean of Pareto solutions in 30 runs. In this work, the MOPSO obtained a mean of 78 solutions and the MGPSO obtained a mean of 103 solutions in Pareto front.

The metric of spacing (S) [9, 8] gives an indication of how evenly the solutions are distributed along the discovered front. The spacing of Pareto front (mean of 30 runs) of MOPSO was 2.7887. On the other hand, the spacing of MGPSO was 1.7479. In terms of spacing, the MGPSO maintains a relatively good spacing metric and obtained a better slightly distribution

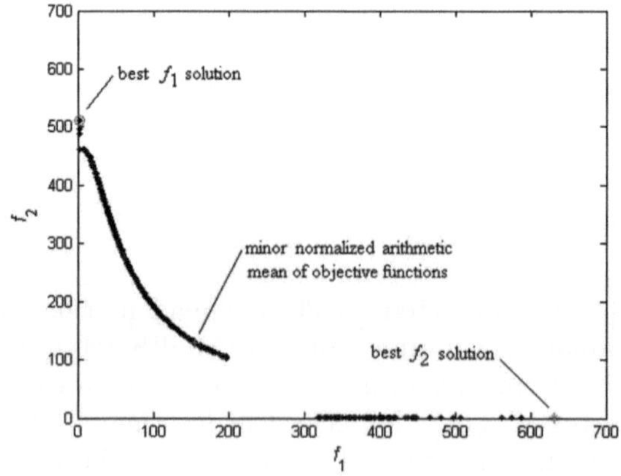

Fig. 1.2. Pareto front of MOPSO obtained from the first simulation

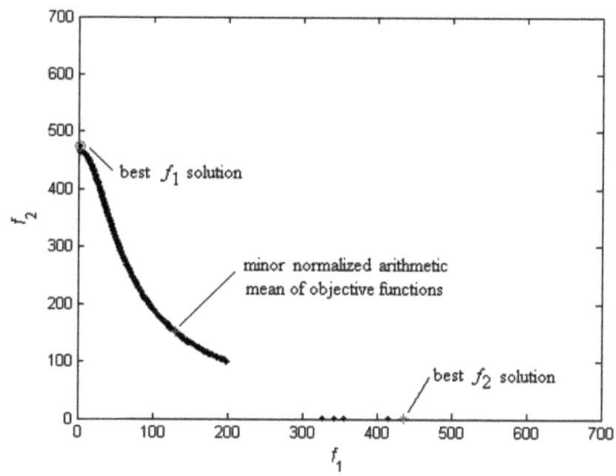

Fig. 1.3. Pareto front of MOPSO obtained from the second simulation

that the MOPSO of non-dominated solutions in Pareto front. Furthermore, the mean distance until the origin of cartesian axis (f_1, f_2) for the Pareto Front in 30 runs is 308.3565 for the MOPSO and 295.5704 for the MGPSO.

Table 1.2 summarizes the results obtained by MOSPO and MGPSO in multi-loop PI tuning for the quadruple-tank process (30 runs). In terms of dominance performance in 30 runs, it can be concluded from Table 2 that MGPSO outperforms MOPSO in terms the best f_1, best f_2, and minor arithmetic mean of f_1 and f_2. For the evaluated case of multi-loop PI tuning, the proposed MGPSO algorithm can be considered a competitive algorithm.

Figure 1.4 to Figure 1.9 present the result in closed-loop of optimized PI controllers for the quadruple-tank process using the data presented in Table 1.2. It ca be observed in Figure 1.5 (best f_1 using MGPSO) that the good performance of PI controllers in closed-loop. In this case, the output y_2 present minor oscillatory behavior in relation to MOPSO (Figure 1.4).

Table 1.2. Results of MOPSO and MGPSO in multi-loop PI tuning for the quadruple-tank process

Method	Index	f_1	f_2	K_{p1}	K_{p2}	T_{i_1}	T_{i_2}
MOPSO	Best f_1	2.554	510.089	34.9318	−0.498	2.3684	255.452
MGPSO	Best f_1	2.277	473.612	47.8038	−0.365	2.2715	236.753
MOPSO	Best f_2	631.580	5.31×10^{-7}	−0.205	49.2183	273.596	1.457
MGPSO	Best f_2	436.715	1.20×10^{-7}	−0.391	45.5404	279.355	1.344
MOPSO	Minor arithmetic mean of f_1 and f_2 (value: 283.7709)	155.405	128.366	34.543	−0.481	312.9481	183.201
MGPSO	Minor arithmetic mean of f_1 and f_2 (value: 280.4498)	129.534	150.916	43.543	−0.475	259.880	185.802

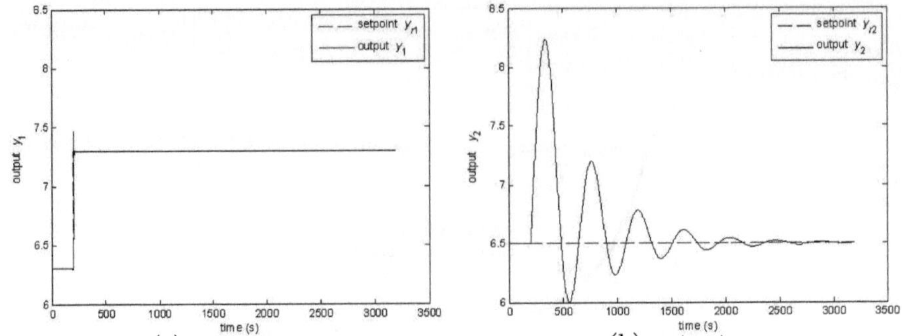

(a) output y_1 (b) output y_2

Fig. 1.4. Best result in terms of f_1 for the MOPSO

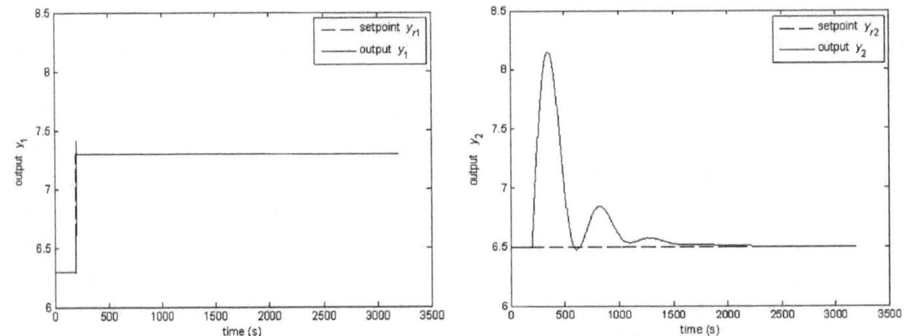

(a) output y_1 (b) output y_2

Fig. 1.5. Best result in terms of f_1 for the MGPSO

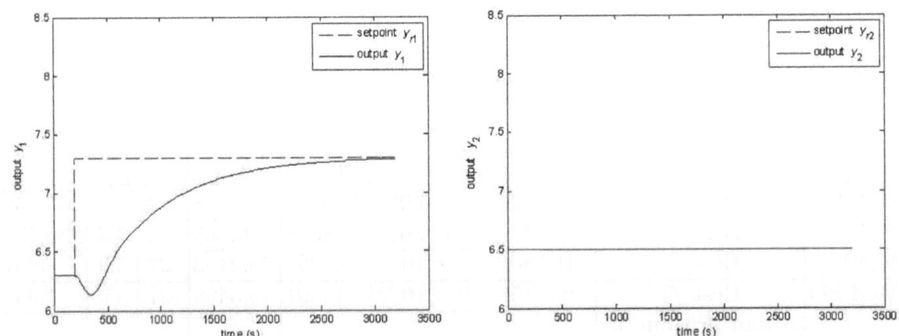

(a) output y_1 (b) output y_2

Fig. 1.6. Best result in terms of f_2 for the MOPSO

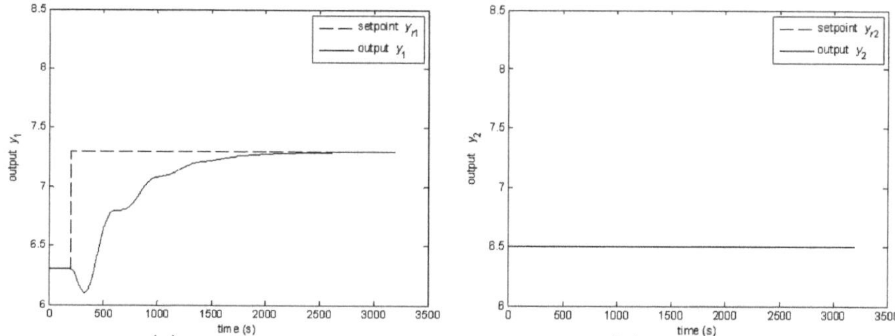

(a) output y_1 (b) output y_2

Fig. 1.7. Best result in terms of f_2 for the MGPSO

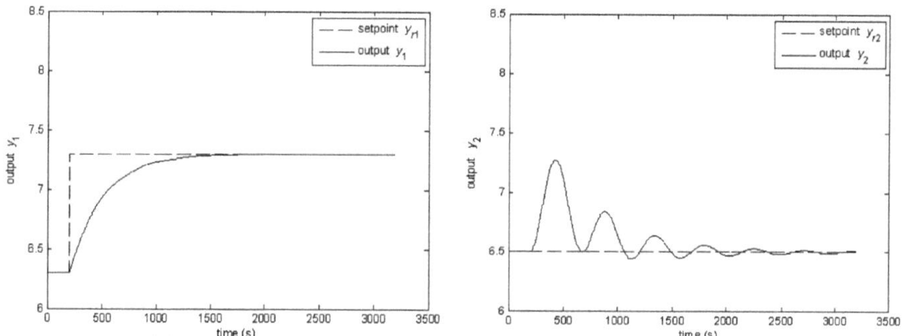

(a) output y_1 (b) output y_2

Fig. 1.8. Best result in terms of arithmetic mean of f_1 and f_2 for the MOPSO

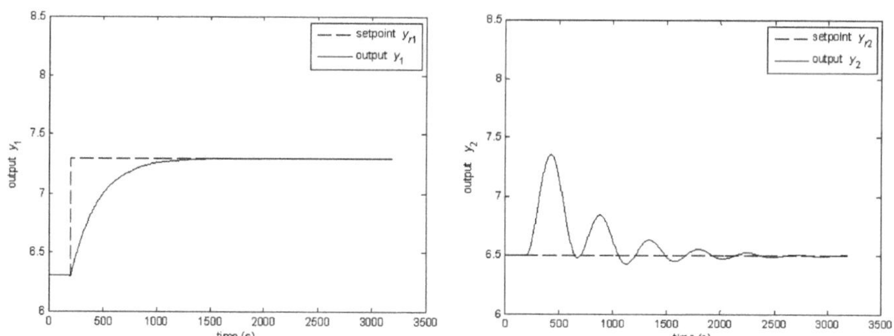

(a) output y_1 (b) output y_2

Fig. 1.9. Best result in terms of arithmetic mean of f_1 and f_2 for the MGPSO

As to the results presented in Figures 1.6 and Figure 1.7, a slow convergence of output y_1 to the proposed set-point can be observed. Wit respect to the results presented in Figures 1.8 and Figure 1.9, the output y_2 presents oscillatory behavior. In the studied case, the closed-loop behavior showed in Figure 1.5, obtained by using MGPSO, gives better results than those presented in Figure 1.8 and Figure 1.9.

1.5 Summary

This chapter presented PSO as a powerful metaheuristic approach inspired by observing the bird flocks and fish schools. Recent works [34, 1] showed that basic PSO algorithm can be modified to accommodate the problem formulation of multiobjective problems, which is to search for a well extended, uniformly distributed, and near-optimal Pareto front.

In this chapter, the MOPSO [31] and the proposed MGPSO design presented promising results to tune the decoupled PI controllers when applied to a quadruple-tank system. The MGPSO allows the discovery of a well-distributed and diverse solution set for PI tuning without compromising the convergence speed of the algorithm. Furthermore, the MOPSO presented competitive results in terms of proximity, diversity, and distribution with the MOPSO for the studied case. The proposed MGPSO method is expected to be extended to other multivariable processes with parameter uncertainties and perturbations. The aim of future works is to investigate the use of MOPSO and MGPSO approaches to tune fuzzy controllers with coupled PID structure.

Acknowledgments

This work was supported by the National Council of Scientific and Technologic Development of Brazil — CNPq, under Grant 309646/2006-5/PQ.

References

1. Abido, M.A.: Multiobjective Particle Swarm Optimization for Environmental/Economic Dispatch Problem. Electric Power Systems Research 79(7), 1105–1113 (2009)
2. Ayala, H.V.H., Coelho, L.S.: A Multiobjective Genetic Algorithm Applied to Multivariable Control Optimization. ABCM Symposium Series in Mechatronics 3, 736–745 (2008)
3. Bobál, V., Böhm, J., Fessl, J., Machácek, J.: Digital Self-Tuning Controllers. Springer, Heidelberg (2005)
4. Carvalho, J.R.H., Ferreira, P.A.V.: Multiple-Criterion Control: A Convex Programming Approach. Automatica 31(7), 1025–1029 (1995)

5. Coelho, L.S., Krohling, R.A.: Predictive Controller Tuning Using Modified Particle Swarm Optimisation Based on Cauchy and Gaussian Distributions. In: Hoffmann, F., Köppen, M., Roy, R. (eds.) Soft Computing: Methodologies and Applications. Springer Engineering Series in Advances in Soft Computing, pp. 287–298. Springer, London (2005)
6. Coello, C.A.C., Pulido, G.T., Lechuga, M.S.: Handling Multiple Objectives with Particle Swarm Optimization. IEEE Transactions on Evolutionary Computation 8(3), 256–279 (2004)
7. Coello, C.A.C.: A Comprehensive Survey of Evolutionary-Based Multiobjective Optimization. Knowledge and Information Systems 1(3), 269–308 (1999)
8. Coello, C.A.C., Van Veldhuizen, D.A., Lamont, G.B.: Evolutionary Algorithms for Solving Multi-Objective Problems. Kluwer Academic Publishers, New York (2002)
9. Deb, K.: Multi-Objective Optimization using Evolutionary Algorithms. Interscience Series in Systems and Optimization. John Wiley & Sons, Chichester (2001)
10. Deb, K., Pratap, A., Agrawal, S., Meyarivan, T.: A Fast and Elitist Multiobjective Genetic Algorithms: NSGA-II. IEEE Transactions on Evolutionary Computation 6(2), 182–197 (2002)
11. Dormido, S., Esquembre, F.: The Quadruple-Tank Process: An Interactive Tool for Control Education. In: Proceedings of European Control Conference, Cambridge, UK (2003)
12. Eberhart, R.C., Kennedy, J.F.: A New Optimizer Using Particle Swarm Theory. In: Proceedings of International Symposium on Micro Machine and Human Science, Japan, pp. 39–43 (1995)
13. Engelbrecht, A.P.: Fundamentals of computational swarm intelligence. John Wiley & Sons, Chichester (2006)
14. Gatzke, E.P., Meadows, E.S., Wang, C., Doyle III, F.J.: Model Based Control of a Four-Tank System. Computers and Chemical Engineering 24(2), 1503–1509 (2000)
15. Higashi, N., Iba, H.: Particle Swarm Optimization with Gaussian Mutation. In: Proceedings of the IEEE Swarm Intelligence Symposium, Indianapolis, IN, USA, pp. 72–79 (2003)
16. Hwang, C.L., Masud, A.S.M.: Multiple Objective Decision Making Methods and Applications: A State of the Art Survey. Springer, Heidelberg (1979)
17. Johansson, K.H.: The Quadruple-Tank Process: A Multivariable Laboratory Process with an Adjustable Zero. IEEE Transactions on Control Systems Magazine 8(3), 456–465 (2000)
18. Kennedy, J.F., Eberhart, R.C.: Particle Swarm Optimization. In: Proceedings of IEEE International Conference on Neural Networks, Perth, Australia, pp. 1942–1948 (1995)
19. Kennedy, J.F., Eberhart, R.C., Shi, Y.: Swarm Intelligence. Morgan Kaufmann Pub, San Francisco (2001)
20. Khan, M.K., Spurgeon, S.K.: Robust MIMO Water Level Control in Interconnected Twin-Tanks Using Second Order Sliding Mode Control. Control Engineering Practice 14(4), 375–386 (2006)
21. Krohling, R.A.: Gaussian Swarm: a novel particle swarm optimization algorithm. In: Proceedings of the IEEE Conference on Cybernetics and Intelligent Systems (CIS), Singapore, pp. 372–376 (2004)

22. Krohling, R.A., Coelho, L.S.: Coevolutionary Particle Swarm Optimization Using Gaussian Distribution for Solving Constrained Optimization Problems. IEEE Transactions on Systems, Man, and Cybernetics — Part B: Cybernetics 36(6), 1407–1416 (2006)
23. Liao, L.-Z., Li, D.: Adaptive Differential Dynamic Programming for Multiobjective Optimal Control. Automatica 38, 1003–1015 (2002)
24. Liu, D., Tan, K.C., Goh, C.K., Ho, W.K.: A Multiobjective Memetic Algorithm Based on Particle Swarm Optimization. IEEE Transactions on Systems, Man, and Cybernetics p– Part B: Cybernetics 37(1), 42–50 (2007)
25. Liu, W., Wang, G.: Auto-Tuning Procedure for Model-Based Predictive Controller. In: Proceedings of IEEE International Conference on Systems, Man, and Cybernetics, Nashville, Tennessee, USA, vol. 5, pp. 3421–3426 (2000)
26. Lu, H., Yen, G.G.: Rank-density-based Multiobjective Genetic Algorithm and Benchmark Test Function Study. IEEE Transactions on Evolutionary Computation 7(4), 325–343 (2003)
27. Moore, J., Chapman, R.: Application of Particle Swarm to Multiobjective Optimization. Department of Computer Science and Software Engineering, Auburn University, Alabama (1999)
28. Nedjah, N., Mourelle, L.M. (eds.): Systems Engineering Using Particle Swarm Optimization. Nova Science Publishers, Hauppauge (2006)
29. Pan, H., Wong, H., Kapila, V., de Queiroz, M.S.: Experimental Validation of a Nonlinear Backstepping Liquid Level Controller for a State Coupled Two Tank System. Control Engineering Practice 13(1), 27–40 (2005)
30. Panda, S.: Multi-objective Evolutionary Algorithm for SSSC-based Controller Design. Electric Power Systems Research 79(6), 937–944 (2009)
31. Raquel, C.R., Naval Jr, P.C.: An Effective Use of Crowding Distance in Multiobjective Particle Swarm Optimization. In: Proceedings of Genetic and Evolutionary Computation Conference (GECCO 2005), Washington, DC, USA (2005)
32. Ratnaweera, A., Halgamuge, S.K., Watson, H.C.: Self-organizing Hierarchical Particle Swarm Optimizer with Time Varying Acceleration Coefficients. IEEE Transactions on Evolutionary Computation 8(3), 240–255 (2004)
33. Secrest, B.R., Lamont, G.B.: Visualizing Particle Swarm Optimization — Gaussian Particle Swarm Optimization. In: Proceedings of the IEEE Swarm Intelligence Symposium, Indianapolis, IN, USA, pp. 198–204 (2003)
34. Sierra, M.R., Coello, C.A.C.: Multi-objective Particle Swarm Optimizers: A Survey of the State-of-the-art. International Journal of Computational Intelligence Research 2(3), 287–308 (2006)
35. Song, M.P., Gu, G.C.: Research on Particle Swarm Optimization: A Review. In: Proceedings of the 3rd International Conference on Machine Learning and Cybernetics, Shanghai, China, pp. 2236–2241 (2004)
36. Van Veldhuizen, D.A., Lamont, G.B.: Multiobjective Evolutionary Algorithms: Analyzing the State-of-the-Art. Evolutionary Computation 8(2), 125–147 (2000)
37. Zambrano, D., Camacho, E.F.: Application of MPC with Multiple Objective for a Solar Refrigeration Plant. In: Proceedings of the IEEE International Conference on Control Applications, Glasgow, Scotland, UK, pp. 1230–1235 (2002)
38. Zitzler, E., Thiele, L., Laumanns, M., Fonseca, C.M., Fonseca, V.G.: Performance Assessment of Multiobjective Optimizers: An Analysis and Review. IEEE Transactions on Evolutionary Computation 7(2), 117–132 (2003)

2

A Non-ordered Rule Induction Algorithm through Multi-Objective Particle Swarm Optimization: Issues and Applications

André B. de Carvalho, Aurora Pozo, and Silvia Vergilio

Computer Sciences Department, Federal University of Paraná,
Curitiba PR CEP: 19081, Brazil
{andrebc,aurora,silvia}@inf.ufpr.br

Multi-Objective Metaheuristics permit to conceive a complete novel approach to induce classifiers, where the properties of the rules can be expressed in different objectives, and then the algorithm finds these rules in an unique run by exploring Pareto dominance concepts. Furthermore, these rules can be used as an unordered classifier, in this way, the rules are more intuitive and easier to understand because they can be interpreted independently one of the other. The quality of the learned rules is not affected during the learning process because the dataset is not modified, as in traditional rule induction approaches. With this philosophy, this chapter describes a Multi-Objective Particle Swarm Optimization (MOPSO) algorithm. One reason to choose the Particle Swarm Optimization Meta heuristic is its recognized ability to work in numerical domains. This propriety allows the described algorithm deals with both numerical and discrete attributes. The algorithm is evaluated by using the area under ROC curve and, by comparing the performance of the induced classifiers with other ones obtained with well known rule induction algorithms. The produced Pareto Front coverage of the algorithm is also analyzed following a Multi-Objective methodology. In addition to this, some application results in the Software Engineering domain are described, more specifically in the context of software testing. Software testing is a fundamental Software Engineering activity for quality assurance that is traditionally very expensive. The algorithm is used to induce rules for fault-prediction that can help to reduce testing efforts. The empirical evaluation and the comparison show the effectiveness and scalability of this new approach.

2.1 Introduction

There is a significant demand for techniques and tools to intelligently assist humans in the task of analyzing very large collections of data, searching

N. Nedjah et al. (Eds.): Multi-Objective Swarm Intelligent Systems, SCI 261, pp. 17–44.
springerlink.com

useful knowledge. Because of this, the area of data mining has received special attention. Data mining is the overall process of extracting knowledge from data. In the study of how to represent knowledge in data mining context, rules are one of the most used representation form. This is because of their simplicity, intuitive aspect, modularity, and because they can be obtained directly from a dataset [21]. Therefore, rules induction has been established as a fundamental component of many data mining systems. Furthermore, it was the first machine learning technique to become part of successful commercial data mining applications [13].

Although many techniques have been proposed and successfully implemented, few works take into account the importance of the comprehensibility aspect of the generated models and rules. Considering this fact, this chapter describes an algorithm, named MOPSO-N (Multi-Objective Particle Swarm Optimization-N) that takes advantage of the inherent benefits of the Particle Swarm Optimization (PSO) technique and puts them in a data mining context to obtain comprehensible, accurate classifiers in the form of simple if-then rules. PSO algorithms are specifically designed to provide robust and scalable solutions. They are inspired on animal swarm intelligence. The particles use simple local rules to govern their actions and via the interactions of the entire group, the swarm achieves its objectives. A type of self-organization emerges from the collection of the group actions. MOPSO-N is an algorithm to induce classifiers aiming to tackle three challenges in this area detailed next.

First, the traditional approach against the multi-objective approach is addressed. Traditional rule induction systems often use a covered approach where a search procedure is iteratively executed. In this search, on each iteration, the algorithm finds the best rule and removes all the examples covered by the rule from the dataset. The process is repeated with the remaining examples [44], and continues until all the examples are covered or some stop criterion is reached. In this way, on each iteration, a new rule is found. However, this approach has major problems. The removal of the examples from the dataset at each new discovered rule causes the over-specialization of the rules after some iteration. This means that each rule covers few examples. Besides that, the classifier composed by the learned rules is an ordered list where the interpretation of one rule depends on the precedent rules.

Despite these traditional algorithms, there are new approaches for rule induction. These approaches are based on Multi-Objective Metaheuristic (MOMH) techniques and two algorithms should be mentioned MOPSO-D [56] that is based on Multi-Objective Particle Swarm Optimization and GRASP-PR [29], based on GRASP with Path-Relinking algorithm. MOMH techniques permit to conceive a novel approach where the properties of the rules can be expressed in different objectives and to find these rules in a unique run. These techniques allow the creation of classifiers composed by rules with specific properties exploring Pareto dominance concepts. These rules can be used as an unordered classifier. The rules are more intuitive and easier to understand because they can be interpreted independently one of the other. Furthermore, the quality of the learned rules is not affected during the learning process because the dataset is not modified.

The second challenge is related to the kind of attributes. Most of the rule induction algorithms, including the algorithms mentioned above, only deal with discrete attributes and, in this way to be used in a certain domain with continuous attributes a previous discretization is necessary. One important feature of the MOPSO-N algorithm, herein described, is its ability to handle with both numerical and discrete attributes. The Particle Swarm Optimization technique has been chosen because this technique presents good results in numerical domains. Differently of MOPSO with only discrete data [56], in our proposed algorithm, for each numerical attribute, the rule learning algorithm tries to discover the best range of values for certain class.

The third challenge addressed in this work is to produce a classifier with good performance in terms of the area under the ROC (Receiver Operating Characteristics) curve. The area under the ROC curve, or simply AUC, has been traditionally used in medical diagnosis since the 1970s, and in recent years has been used increasingly in machine learning and data mining research. So, to tackle this purpose we choose two objectives, the sensitivity and specificity criteria that are directly related with the ROC curve. The sensitivity is equivalent to Y axis of ROC graph and the specificity is the complement of the X axis. The hypothesis behind these strategies is: classifiers composed by rules that maximize these objectives present good performance in terms of AUC.

This work extends a previous paper [16], where some of the idea and preliminary results are reported. Here the foundations of the algorithm and operators proposed are discussed in more detail and a wider set of experiments is presented.

To validate MOPSO-N its results are compared with other known techniques from the literature. All algorithms are applied to different datasets and the results are compared using AUC. Furthermore, the analysis of the Pareto Front coverage of MOPSO-N and MOPSO-D systems is performed.

Finally, this chapter describes the application of MOPSO-N in the Software Engineering domain. Software plays a crucial role in all most areas and human activities. Nowadays, tasks for quality assurance, such as software testing can be considered fundamental in the Software Engineering area. However, this activity traditionally consumes a lot of effort and time. For almost systems, the number of statements, methods or classes to be tested can be very large, and the number of paths in a program is usually infinite. To reduce these limitations and the test costs, a good strategy is to select some classes with a greater probability of containing a fault and to focus testing efforts on these classes. MOPSO-N is used to induce rules for fault-prediction that can help to reduce testing efforts. The empirical evaluation and the comparison show the effectiveness and scalability of this new approach.

The rest of this chapter is organized as follows. Section 2.2 reviews rule learning concepts. Section 2.3 discusses the performance assessment methodology used in this work. Section 2.4 presents the main multiple objective particle swarm aspects. Section 2.5 describes the proposed algorithm. Section 2.6 explains the empirical study performed to evaluate the algorithm. Section 2.7 describes the application of MOPSO-N to prediction of faulty classes. Section 2.8 presents a discussion of related works. Finally, Section 2.9 concludes the paper and discusses future works.

2.2 Rule Learning Concepts

The algorithm described in this chapter is applied to the learning rules problem. A rule is a pair <antecedent, consequent> or if *antecedent* then *consequent* where both the antecedent and the consequent are constructed from the set of attributes.

Let Q be a finite set of attributes, which in practice correspond to the fields in the database. Each $q \in Q$ has an associated domain, $Dom(q)$. An attribute test, b, consists of an attribute, $at(b) \in Q$, and a value set, $Val(b) \in Dom(at(b))$, and may be written $at(b) \in Val(b)$. A record satisfies this test if its value for attribute $at(b)$ belongs in the set $Val(b)$. An algorithm may allow only certain types of value sets $Val(b)$. Types of categorical attribute tests are as follows:

- Value: $Val(b) = \{v(b)\}$, where $v(b) \in Dom(at(b))$. This may be written $at(b) = v(b)$.
- Inequality: $Val(b) = \{x \in Dom(at(b)) : x \neq v(b)\}$, where $v(b) \in Dom(at(b))$. This may be written $at(b) \neq v(b)$.
- Subset: Val(b) unrestricted, i.e. any subset of Dom(at(b)).

Types of numerical attribute tests are as follows:

- Binary partition: $Val(b) = \{x \in Dom(at(b)) : x \leq v(b)\}$ or $Val(b) = \{x \in Dom(at(b)) : x \geq v(b)\}$, where $v(b) \in Dom(at(b))$. In this case, an attribute test may be written $at(b) \leq v(b)$ or $at(b) \geq v(b)$, respectively.
- Range: $Val(b) = \{x \in Dom(at(b)) : l(b) \leq x \leq u(b)\}$, where $l(b), u(b) \in Dom(at(b))$. Here, the AT is written $l(b) \leq at(b) \leq u(b)$.

Each rule created has the same consequent, consisting of just one AT (c_k). This defines the class of interest. We say that the rule r_k covers the example e_i if the example satisfies the antecedent of r_k. And, if the rule r_k covers e_i and $c_k = c_i$ then the rule correctly classifies the example e_i, otherwise, there is an incorrect classification.

The number of examples correctly and incorrectly classified are used to calculate the measures for rules evaluation, and are resumed in the contingency table. Table 2.1 shows the contingency table, where B denotes the set of instances for which the body of the rule is true, i.e., the set of examples covered by the rule r_k, and $\overline{B}$ denotes its complement (the set of instances not covered by the rule r_k). H is the set of instances where consequent is true ($c_i = c_k$) and $\overline{H}$ is its complement. HB then denotes $H \cap B$, $\overline{H}B$ denotes $\overline{H} \cap B$, and so on.

Table 2.1. A Contingency Table

	B	$\overline{B}$	
H	$n(HB)$	$n(H\overline{B})$	$n(H)$
$\overline{H}$	$n(\overline{H}B)$	$n(\overline{H}\overline{B})$	$n(\overline{H})$
	$n(B)$	$n(\overline{B})$	N

Where $n(X)$ denotes the cardinality of the set X, e.g., $n(\overline{H}B)$ is the number of instances for which H is false and B is true (i.e., the number of instances erroneously covered by the rule). N denotes the total number of instances in the dataset.

From the contingency matrix it is possible to calculate measures such as: True Positive rate (*TP rate*), True Negative rate (*TN rate* or specificity), False Positive rate (*FP rate*) and False Negative rate (*FN rate*). *TP rate*, also called sensitivity, is the precision between the positive examples (Equation 2.1). Its complement is the *FN rate* (i.e. *FN rate* = *1 - FP rate*). Specificity is the precision between the negative examples (Equation 2.2). Its complement is the *FP rate*.

$$sensitivity = \frac{n(HB)}{n(H)} \tag{2.1}$$

$$specificity = \frac{n(\overline{HB})}{n(\overline{H})} \tag{2.2}$$

Other common rule evaluation measures are: support and confidence. The support is the number of correctly classified examples by the number of examples (Equation 2.3); and confidence that is the number of correctly classified examples by the number of covered examples (Equation 2.4). A review of the metrics used for rules evaluation is presented in [36].

$$support = \frac{n(HB)}{N} \tag{2.3}$$

$$confidence = \frac{n(HB)}{n(B)} \tag{2.4}$$

2.2.1 AUC

Rules are usually aggregated into a rule set for the purpose of classification [21]. For several years, the most used performance measure for classifiers was the accuracy [3]. The accuracy is the fraction of examples correctly classified, showed on Equation 2.5. Despite of its use, the accuracy maximization is not an appropriate goal for many of the real-world tasks [47]. A tacit assumption in the use of classification accuracy as an evaluation metric is that the class distribution among examples is constant and relatively balanced. In real world this case is rare, moreover, the cost associated with the incorrect classification of each class can be different because some classifications can lead to actions which could have serious consequences [48].

$$accuracy = \frac{n(HB) + n(\overline{HB})}{N} \tag{2.5}$$

$$TPrate = \frac{n(HB)}{n(H)} \tag{2.6}$$

$$FPrate = \frac{n(H\overline{B})}{n(\overline{H})} \tag{2.7}$$

Based on the explanation above, many people have preferred the Receiver Operating Characteristic (ROC) analysis. The ROC graph has been used in signal detection theory to depict tradeoffs between TP rate (Equation 2.6) and FP rate (Equation 2.7) [19]. However, it was only introduced in Machine Learning context at the end of 90's [48]. The ROC graph has two dimensions: the TP rate at the y-axis and the FP rate at x-axis.

The best classifiers have high TP rate and low FP rate. The most northwest the point, the best will be the associated classifier. In Fig. 2.1, the classifier associated to Point B is better than the classifier of Point C. We can say that Classifier B dominates Classifier C. Otherwise, we can not say the same about Classifiers A and C, because Point A has a lower value for the FP rate, however it has also a lower value for TP rate. For this case, Point A is not dominated by Point C, neither C is dominated by A.

If one classifier is not dominated by any other, this classifier is called as a non-dominated one. That is, there is no other classifier with a higher TP rate and a lower FP rate. The set of all non-dominated classifiers is known as Pareto Front. The continuous line of Fig. 2.1 shows the Pareto Front.

The ROC curve gives a good visualization of the performance of a classifier [8]. However, frequently, for comparison purposes of learning algorithms, a single measure for the classifier performance is needed. In this sense, the Area Under the ROC Curve (AUC) was proposed [23] [50] [52]. The AUC measure has an important statistical property: it is equivalent to the probability that a randomly chosen positive instance will be rated higher than a negative instance and thereby is also estimated by the Wilcoxon test of ranks [25]. The AUC values for non-ordered set of rules are calculated using a weighted voted classification process [21].

Given a rule set containing rules for each class, we use the best k rules of each class for prediction, with the following procedure: (1) select all the

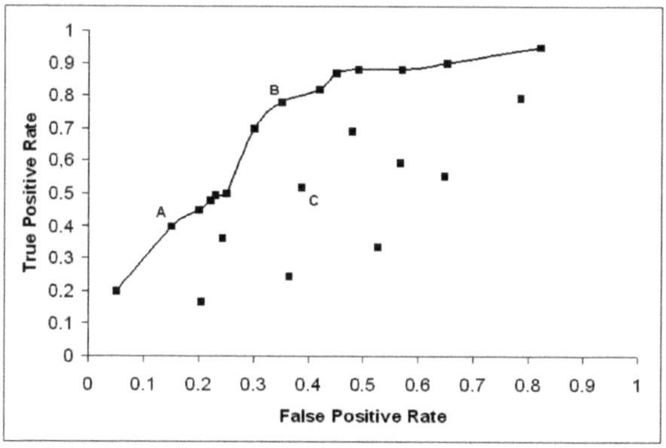

Fig. 2.1. A ROC Graph plotted with a set of classifiers. The continuous line shows the Pareto Front.

rules whose bodies are satisfied by the example; (2) from the rules selected in Step 1, select the best k rules for each class; and (3) compare the expected average of the positive confidence of the best k rules of each class and, choose the class with the highest expected positive confidence as the predicted class. After the vote of all rules, each example has an associated value that is the sum of confidence of positive rules less the sum of confidence of negative rules. Thus, for each instance of the database, we obtain a numerical rank with the associated values. This rank can be used as a threshold to produce a binary classifier. If the rank of the instance goes beyond the threshold, the classifier produces a "yes", otherwise, a "no". Each threshold value generates a different point in the ROC plane. So, varying the threshold from $-\infty$ to $+\infty$ one produces a curve on the ROC plane and we are able to calculate the area under curve (AUC) [46]. This value gives a good visualization of the classifier performance [8]. For a perfect classification, the positive examples have the greatest values grouped at the top of the ranking; and AUC is equal to 1 (its greatest value).

For maximizing the AUC value, it is preferable that positive examples receive more votes from positive rules than from negative ones, i.e., it is expected: high precision between the positive examples (sensitivity) and high precision between the negatives (specificity). For this reason, it is expected that the Pareto Front with sensitivity and specificity criteria maximizes AUC.

2.3 Performance Assessment of Stochastic Multi-Objective Optimizers

In the last years, there has been a growing interest in applying stochastic search algorithms such as evolutionary ones to approximate the set of Pareto optimal solutions and as a consequence the issue of performance assessment has become more important. This section revises some definitions and concepts associated to Multi-Objective Optimizers. Optimization problems that have more than one objective function are called Multi-Objective problems. In such problems, the objectives to be optimized are usually in conflict in respect to each other, which means that there is not a single solution for these problems. Instead, the goal is to find a good "trade-off" solutions that represent the better possible compromise among the objectives. The general multi-objective maximization problem (with no restrictions) can be stated as to maximize Equation 2.8.

$$f(x) = (f_1(x), ..., f_Q(x)) \tag{2.8}$$

subjected to $x \in X$, where: x is a vector of decision variables and X is a finite set of feasible solutions. Function $f(x)$ maps the set of feasible solutions $X \in Z$ to the Q-dimensional objective space, $Q > 1$ being the number of objectives. Then, $f : X \rightarrow Z$ is a function that assigns an objective vector $z = f(x) \in Z$ to each solution $x \in X$. Let $z^1 = (z_1^1, ..., z_Q^1)$ and $z^2 = (z_1^2, ..., z_Q^2)$, $z^1, z^2 \in Z$ be two objective vectors. Some dominance definitions are as follows.

- $z^1 \succ z^2$ (z^1 dominates z^2) if z^1 is not worse than z^2 for any objective and is better in at least one
- $z^1 \succ\succ z^2$ (z^1 strictly dominates z^2) if z^1 is better than z^2 for all objectives
- $z^1 \geq z^2$ (z^1 weakly dominates z^2) if z^1 is not worse than z^2 for any objective
- $z^1 \mid\mid z^2$ (z^1 and z^2 are incomparable to each other) if neither z^1 dominates z^2 nor z^2 dominates z^1
- $z^1 \sim z^2$ (z^1 and z^2 are indifferent) if z^1 and z^2 are equal for all objectives.

The goal is to discover solutions that are not dominated by any other in the objective space. A set of non-dominated objective vectors is called Pareto optimal and the set of all non-dominated vectors is called Pareto Front. The Pareto optimal set is helpful for real problem, e.g., engineering problems, and provides valuable information about the underlying problem [35]. In most applications, the search for the Pareto optimal is NP-hard [35], then the optimization problem focuses on finding a set as close as possible to the Pareto optimal, an approximation set.

Let $A \subseteq Z$ be a set of objective vectors. A is said to be an approximation set if any two elements of A are incomparable to each other. The dominance relations can be extended to approximation sets. Given two approximation sets A_1 and A_2, A_1 is said to dominate A_2 ($A1 \succ A2$) if every solution vector of A_2 is dominated by at least one objective vector of A_1. The other relations are defined accordingly. In order to compare algorithms, it is also useful to define the relation *is better than*. It is said that an approximation set A_1 *is better than* other, A_2, ($A_1 \rhd A_2$) if $A_1 \geq A_2$ and $A_1 \neq A_2$. A_1 and A_2 are said to be incomparable to each other ($A_1 || A_2$) if neither $A_1 \geq A_2$ nor

$A_2 \geq A_1$. Those definitions are very useful when comparing the outcomes of approximation algorithms.

As example, Fig. 2.2 shows the Pareto Front associated to the "all rules" of dataset 2 (*ecoli*) of Table 2.2, when sensitivity and specificity are used as objectives.

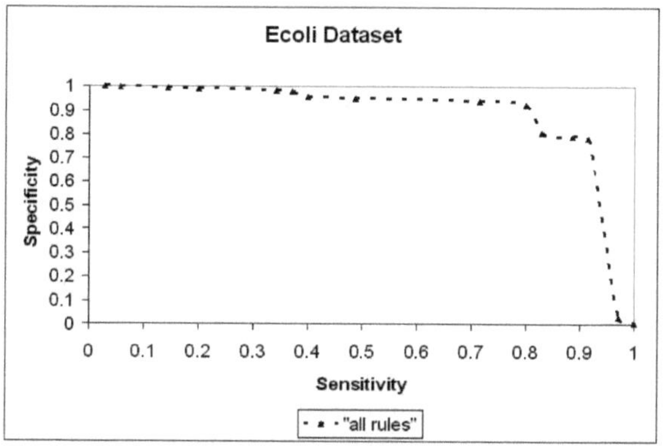

Fig. 2.2. Pareto Front of the positive rules for Dataset 2 (*ecoli*)

One way to evaluate the stochastic multiobjective problems is to compare the approximation sets samples. Fig. 2.3 shows the approximation sets of 30 independent executions of MOPSO-D algorithm and 30 independent executions of MOPSO-N algorithm for Data Set 2 (*ecoli*). Two approaches can be used to compare the approximations sets: dominance ranking and quality indicators [35].

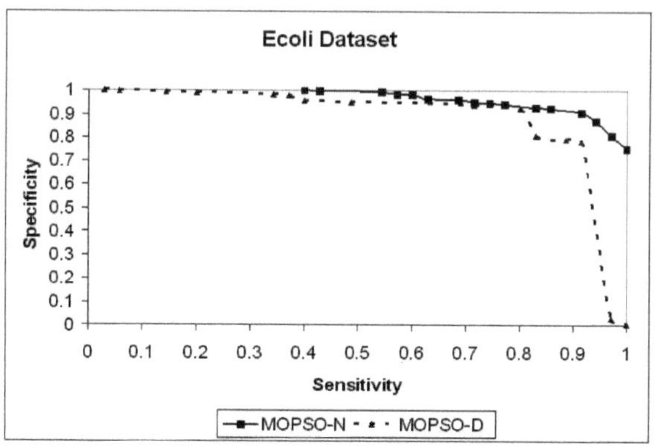

Fig. 2.3. Approximation sets of the positive rules for Dataset 2(*ecoli*) generated by both MOPSO algorithms

2.3.1 Performance Assessment

The general directions for the performance assessment methodology used in this work are given at [35]. This work recommends to use the complementary strengths of three approaches. As a first step in a comparison, any significant differences between the optimizers considered should be probed using the dominance-ranking approach, because such analysis allows the strongest type of statements to be made. After, quality indicators can then be applied in order to quantify the potential differences in quality and to detect differences that could not be revealed by dominance raking. Finally, the computation and the statistical comparison of the empirical attainment functions. These approaches are better explained below.

Dominance ranking stands for a general, preference independent assessment method that is based on pairwise comparisons of approximation sets. The dominance ranking can be computed using the additive binary epsilon indicator [59] and the Mann-Whitney test [15]. Given two approximation sets, A and B, the binary epsilon indicator, $I_\xi(A, B)$ gives the minimum factor by which each point of B can be added in such manner that the resulting set is weakly dominated by A. Given two sets of stochastic algorithms executions $A = \{A_1, ..., A_k\}$ and $B = \{B_1, ..., B_j\}$, a rank is given for each approximation set $Ci \in C$, $C = A \cup B$, being the number of approximation sets weakly dominating C_i plus one (Equation 2.9). Thus, the smaller the rank an approximation set C_i receives the better it is.

$$rank(C_i) = 1 + |\{C_j \in C : C_j \rhd C_i\}| \tag{2.9}$$

Equation 2.9 provides an integer value that is assigned to each approximation set. Then, a statistical test can be performed in order to establish whether significant differences exist between sets A and B. In particular, we are interested in discover if the ranks assigned to the approximation sets of one algorithm are significantly smaller than the ranks assigned to the approximation sets of the other one. The Mann-Withney U-test, also called Mann-Whitney-Wilcoxon test or Wilcoxon rank-sum test, is a non-parametric test used to verify the null hypothesis that two samples come from the same population [15]. If the results obtained with the dominance ranking test, however, do not support conclusions about significant differences among the investigated algorithms, then new experiments have to be done.

Quality indicators represent a means to express and measure quality differences between approximation sets, on the basis of additional preference information. In this work, unary quality indicators were used. The unary quality indicator is a function $I(X)$ that maps one approximation set X into a real value. Let A and B be a pair of approximation sets. The difference between their corresponding indicator values $I(A)$ and $I(B)$ reveals a difference in the quality of the two sets. The unary quality indicators hypervolume [59], epsilon [60] and $R2$ [26] are the most used.

The hypervolume indicator $I_H(A)$ measures the hypervolume of that portion of the objective space that is weakly dominated by an approximation set

A [59]. In order to utilize this indicator, the objective space must be bounded. If it is not, a reference point R (dominated by all points) must be defined. The hypervolume difference to a reference set R can also be considered, if the hypervolume difference to a set R is used, then smaller values correspond to higher quality of the approximation set A. The hypervolume indicator is the only unary indicator that is capable to detect that one approximation set A is not better than another approximation set B for all pair of solutions [35].

The unary epsilon additive $I_\epsilon^1(A)$ gives the minimum factor ϵ by which each point in R can be added in such way that the resulting transformed approximation set is weakly dominated by A [60]. One approximation set A is preferred to another B, according to the unary epsilon additive indicator, if $I_\epsilon^1(A) < I_\epsilon^1(B)$. If the hypervolume and the unary epsilon additive indicate opposite preferences for two approximation sets then they are incomparable. The $R2$ indicator, I_{R2} [26] is based on a set of utility functions. A utility function is a mapping of the set Z of q-dimensional objective vectors to the set of real numbers.

Equation 2.10 presents the augmented Tchebycheff function that was utilized in the I_{R2} indicator used in this paper. In this equation z_j^* is a point that weakly dominates the point z_j, $\xi = 0.01$ and $\lambda_j \in \Delta$, the set of weight vectors containing $rank(C_i)$ (Equation 2.9) uniformly dispersed scalarizing vectors.

$$u_\lambda = -\left(\max_{j \in 1..q} \lambda_j \left| z_j^* - z_j \right| + \xi \sum_{j=1}^{n} \left| z_j^* - z_j \right| \right) \tag{2.10}$$

The same reference set is employed for those indicators and, when necessary, the same reference points. The reference set is formed by the non dominated vectors of all approximation sets generated by those algorithms. The Kruskal-Wallis statistical test [15] is used to compare the algorithms based on those three quality indicators. The Kruskal-Wallis test is a logical extension of the Mann-Whitney test. It is also a non-parametric test used to compare three or more samples testing the null hypothesis that all populations have identical distribution functions. For both algorithms under consideration, new independent data is generated for the statistical test, with independent runs executed for each instance.

The reference points (the best and the worst) of each data set are determined by analyzing the points in the reference set. The best and the worst values of each objective are calculated. The reference point for the hypervolume indicator is formed by the worst values of each objective plus 10. The reference point $z*$ of the $R2$ indicator is obtained with the best value of each objective.

2.4 Multi-Objective Particle Swarm Optimization

Particle Swarm Optimization (PSO), developed by Kennedy and Eberhart [33], is a population-based heuristic inspired by the social behavior of bird flocking aiming to find food. PSO have some similarities with evolutionary algorithms: both systems are initialized with a set of solutions, possibly random, and search for optima by updating generations. Despite the fact that they share some similarities, there are two main differences between them. First, there is no notion of offspring in PSO, the search is guided by the use of leaders. Secondly, PSO has no evolution operators such as crossover or mutation. In Particle Swarm Optimization, the set of possible solutions is a set of particles, called swarms moving in the search space, in a cooperative search procedure. These moves are performed by an operator that is guided by a local and a social component [34]. This operator is called velocity of a particle and moves it through an n-dimensional space based on the best positions of their neighbors (social component) and, on their own best position (local component). The best particles are found based on the fitness function, which is the problem objective function.

Each particle p_i, at a time step t, has a position $x(t) \in R^n$, that represents a possible solution. The position of the particle, at time $t+1$, is obtained by adding its velocity, $v(t) \in R^n$, to $x(t)$:

$$\vec{x}(t+1) = \vec{x}(t) + \vec{v}(t+1) \tag{2.11}$$

The velocity of a particle p_i is based on the best position already fetched by the particle, $p_{best}(t)$, and the best position already fetched by the neighbors of p_i, $g_{best}(t)$. This neighborhood could be one single particle or the whole population in analysis. The velocity update function, in time step $t+1$ is defined:

$$\vec{v}(t+1) = \varpi \times \vec{v}(t) + \phi_1 \times (\vec{p}_{best}(t) - \vec{x}(t)) + \phi_2 \times (\vec{g}_{best}(t) - \vec{x}(t)) \tag{2.12}$$

The variables ϕ_1 and $\phi 2$, in Equation 2.12, are coefficients that determine the influence of the particle best position, $p_{best}(t)$, and the particle global best position, $g_{best}(t)$. The coefficient ϖ is the inertia of the particle, and controls how much the previous velocity affects the current one. After the velocities and positions of all the particles have been updated, $p_{best}(t+1)$ and $g_{best}(t+1)$ are calculated. This process continues to the next generation, until the execution is ended.

To define if one solution is better than other, the fitness function is used. As each particle's position represents a solution to the problem, the fitness of a particle p, represented as $\alpha(x)$, is a function of the particle p position, $\alpha : S \subseteq R^n \rightarrow R$. So, for a minimization problem, at time t, a particle p_i with a position x_i, is better than a particle p_j with position x_j, if:

$$\alpha(\vec{x_i}(t)) < \alpha(\vec{x_j}(t))$$

The inverse is true for maximization problems. In multi-objective problems, there is more than one fitness function. So the best particles are found based on the Pareto's dominance concept [43]. With this concept, in most cases, there is no best solution, but a set of non-dominated solutions. Let Π

be the set of possible solutions to the problem. Let $\vec{x} \in \Pi$ and $\vec{y} \in \Pi$ be two solutions in this set, represented as a particle's position. For a minimization problem, the solution $\vec{x}$ dominates $\vec{y}$ if:

$$\forall \alpha \in \Phi, \alpha(\vec{x}) \leq \alpha(\vec{y}), and$$
$$\exists \alpha \in \Phi, \alpha(\vec{x}) < \alpha(\vec{y})$$

where Φ represents the set of all objective functions.

When there is no solution $\vec{z}$ that dominates $\vec{x}$, it said that it is a non-dominated solution. The set of all non-dominated solutions for a problem is the Pareto Front [51].

In Multi-Objective Particle Swarm Optimization, MOPSO, there are many fitness functions. In this way, it is possible to obtain results with specific properties by exploring Paretos dominance concepts. Based on these concepts each particle of the swarm could have different leaders, but only one may be selected to update the velocity. This set of leaders is stored in a repository, which contains the best non-dominated solutions found. At each generation, the velocity of the particles is updated according to the following equation:

$$\vec{v}(t+1) = \varpi \times \vec{v}(t) + \phi_1 \times (\vec{p}_{best}(t) - \vec{x}(t)) + \phi_2 \times (\vec{R_h}(t) - \vec{x}(t)) \quad (2.13)$$

$\vec{R}_h$ is a particle from the repository, chosen as a guide. There are many ways to make this choice, as demonstrated in [51]. At the end of the algorithm, the solutions in the repository are the final output.

One possible way to make the leader's choice is called sigma distance [41]. This method, accordingly to the results presented in [51], is one of the most adequate for the PSO technique. For a two-objective problem, the sigma values is defined in the follow way:

$$\sigma = \frac{f_1(x)^2 - f_2(x)^2}{f_1(x)^2 + f_2(x)^2} \quad (2.14)$$

For problems with more than two objectives, the sigma value is a vector with $\binom{n}{2}$ elements, where n is the number of objectives of the problem. Each element of this vector represents a combination of two elements applied to Equation 2.14. The leader for a particle of the swarm, R_h, is the particle from the repository which has the smallest Euclidian distance between its sigma vector and the sigma vector of the swarm particle.

2.5 Rule Learning with MOPSO

This section explains the rule learning algorithm, MOPSO-N. This algorithm follows the philosophy of the algorithm MOPSO-D [56], with the difference that it is able to deal with both numerical and discrete attributes. In this way, if the dataset has only discrete attributes, MOPSO-N works like MOPSO-D. The main advantage of MOPSO-N is that it does not need a pre-processing step to the data discretization. This step is executed out of the search procedure and could introduce some lost of information. MOPSO-N constructs data ranges for each attribute that produce good classification results.

The algorithm follows the Michigan approach where each particle represents a single solution or a rule. In this context, a particle is an n-dimensional vector of real numbers. One real number represents the value for each discrete attributes and two real numbers represent an interval for numerical attributes. The interval is defined by its lower and upper values. Each attribute can accept the value '?', which means that, for that rule, the attribute does not matter for the classification.

The rule learning algorithm using MOPSO-N (Algorithm 1), works as follows. First, an initial procedure is executed where the position of each particle is initialized and all particles are spread randomly in the search space. The discrete attributes are defined using a roulette procedure, where all the values of the attributes, including the void value, have equal probabilities. For the numerical attributes, first, all attributes have the probability to be empty. This probability is a parameter of the algorithm, denominated `prob`. In the proposed approach, `prob` was set with a low value, to favor the creation of rules with different initial ranges and to explore a greater number of areas of the search space. If an attribute is set as non-empty, the lower and upper limits are spread randomly in the interval defined by the `minimum` and `maximum` values for the attribute obtained from the dataset. After that the velocity of all particles is initialized (Line 3) and the local leader is defined as the initial position (Line 4).

The next step is the evaluation of each particle for all objectives. After the evaluation, the repository is initialized with the non-dominated solutions (Line 6). From this point, we must choose the leaders for each particle of the population. Then, the objective space between the particles in the repository is divided (Line 7). In this approach the sigma distance, discussed in the previous section, is used. After this initial configuration, the evolutionary loop is executed until a stop criterion is reached.

In this work this criterion is the maximum number of generations. In each iteration, all particles are evaluated. The operations discussed in the previous section are implemented. The velocity of the particle is updated (Line 10) and then, the new positions are calculated (Line 12). In those operations, a `mod` operator is applied. This operation is performed to limit the particle into the search space. The `mod` operator was chosen because all the attributes values will have an equal probability to be chosen. In the velocity update process, `mod` is only applied on discrete attributes. In the calculation of the position, the numerical attributes are modified with the `mod` operator. In this case, `mod` is executed with respect the maximum and minimum values of the interval for the attribute. If the new upper limit overflows the maximum value, the excess is added to the minimum value, and that is the new limit. The same procedure is executed for the lower limit, in the inverse way. After

Algorithm 1. Rule learning algorithm with MOPSO-N

for each particle i **do**
  Initialize $\overrightarrow{x_i}$ with a random solution to the problem
  Initialize $\overrightarrow{v_i}$ with a random velocity.
  Initialize $\overrightarrow{pbest_i} = \overrightarrow{x_i}$
end for
Evaluate particles using fitness functions.
Find non-dominated solutions, storing them in the repository.
Divide the search space between the solutions of the repository
while not stop criterion **do**
  for each particle i of the swarm **do**
    $\overrightarrow{v}(t+1) = (\varpi \times \overrightarrow{v}(t) + \phi_1 \times (\overrightarrow{pbest}(t) - \overrightarrow{x}(t)) + \phi_2 \times (\overrightarrow{R_h}(t) - \overrightarrow{x}(t)))mod\overrightarrow{N_i}$
    Note: $\overrightarrow{N_i}$ is a vector of the number of possible values to each attribute of the database. It restricts the particle inside the search space. Applied only for discrete attributes.
    $\overrightarrow{x}(t+1) = (\overrightarrow{x}(t) + \overrightarrow{v}(t+1))mod\overrightarrow{N_i}$
    Note: For numerical attributes, the value of $\overrightarrow{N_i}$ is the attribute range, defined in the dataset. Evaluate particles. The particle will have one value for each objective of the problem.
    Update $\overrightarrow{pbest}(t)$
    Update the repository with non-dominated particles.
    Divide the search space, finding $\overrightarrow{R_h}(t)$ of the particles.
  end for
end while
return Repository

this process, the smaller value is the new lower limit, and the larger is the upper. If both values overflow the limits, the attributes are set to empty ('?').

The final rules learned by MOPSO are the non-dominated solutions contained in the repository, at the end of the process.

2.6 MOPSO Evaluation

To validate the MOPSO algorithm, this section presents two experimental studies. In the first study, the classification results of MOPSO are evaluated. The results of MOPSO are compared with other well known algorithms from the literature, using the AUC measure. The chosen algorithms are C4.5 [49], C4.5 with No Pruning, RIPPER [14] and NNge [40]. The second study analysis the Pareto Front coverage of MOPSO. This analysis measures the quality of the generated rules. Here, the Pareto front of the MOPSO-D is compared with the real Pareto Front obtained through an "all rules" algorithm. After, the fronts of MOPSO-N are compared with the MOPSO-D and an algorithm that generates all possible rules (named here "all rules"). Sections 2.6.1 and 2.6.2 present the details of each study.

2.6.1 AUC Comparison

The classification study considers a set of experiments with 8 databases from the UCI [2] machine learning repository. The datasets are presented on Table 2.2. The chosen algorithms are C4.5, C4.5 No Pruning, RIPPER and NNge, all of them were applied using the tool Weka [24]. The experiments were executed using 10-fold-stratified cross validation and for all algorithms were given the same training and test files. The database with more than two classes were reduced to two-class problems, positive and negative, selecting the class with the lower frequency as positive and joining the remaining examples as negative. All the databases have attributes with numerical values. No pre-processing step was applied.

Table 2.2. Description of the experimental data sets

#	Data set	Attributes	Examples
1	breast	10	683
2	ecoli	8	336
3	flag	29	174
4	glass	10	214
5	haberman	4	306
6	ionosphere	34	351
7	new-thyroid	6	215
8	pima	9	768

MOPSO-N was executed with 100 generations and 500 particles for each class. For each fold was made thirty runs. The parameters ω, ϕ_1 and ϕ_2, are randomly chosen by the algorithms in each update of the velocity of a particle. ω varies in the interval $[0, 0.8]$, and both ϕ_1 and ϕ_2 varies in $[0, 4]$. All these control parameters were experimentally set and were derived from a previous work presented in [56]. A future work, based on some works in the literature [9, 51], will try to discover the best values of the control parameters for the MOPSO-N. The *prob* parameter, as said before, has a low value and was empirically set to 0.1. This value were achieved from a experiment varying the *prob* value from 0 to 1, increment increasing 0.1 for each execution. This procedure did not show great results differences between each *prob* value, but a low value of this parameter was chosen to permit a larger cover of the search space in the beginning of the execution. The AUC values for the MOPSO-N algorithm were calculated by using a confidence voted classification process as explained in Section 2.2.1.

The AUC values are shown in Table 2.3 and the number between brackets indicates the standard deviation. The non-parametric Wilcoxon test, U-Test, with 5% confidence level, was executed to define which algorithm is better. In this test, all the ten AUC fold values of all thirty executions of MOPSO (in a total of three hundred AUC values) were compared with the ten AUC fold values of the other algorithms. Because the algorithms chosen from literature are deterministic, to perform the Wilcoxon test, the AUC values were

Table 2.3. Experiments Results: Mean AUC

Datasets	MOPSO-N	C4.5	C4.5 NP	RIPPER	NNge
1	**98.97 (1.33)**	95.19 (2.54)	95.67 (2.85)	96.98 (2.19)	96.92 (2.26)
2	**78.62 (31.59)**	78.90 (18.83)	79.13 (18.64)	79.51 (11.62)	72.35 (9.36)
3	**81.71 (17.36)**	50.00 (0.0)	54.07 (23.05)	52.75 (15.01)	59.16 (16.29)
4	75.23 (16.36)	**80.94 (17.24)**	**81.69 (17.45)**	54.11 (10.83)	54.50 (15.20)
5	**69.63 (10.72)**	59.59 (8.82)	62.29 (9.16)	61.42 (8.06)	56.14 (6.86)
6	88.53 (6.88)	86.52 (6.28)	88.14 (5.90)	**90.96 (4.68)**	**90.49 (3.87)**
7	**95.94 (9.93)**	91.80 (11.45)	91.89 (11.49)	92.77 (11.55)	92.50 (11.61)
8	**81.30 (4.54)**	75.58 (3.56)	75.48 (4.30)	71.29 (3.88)	68.51 (6.81)
Avg	83.74	77.31	78.54	74.97	73.82

repeated thirty times. In Table 2.3 the cells highlighted indicate which algorithm obtained the best result.

Table 2.3 shows that for almost all datasets (Datasets 1, 2, 3, 5, 7 and 8), MOPSO presents the best result, accordingly the U-Test. Furthermore, in some datasets, like Datasets 3 and 7, the MOPSO-N achieved a very good classification result, when compared with the chosen algorithms. For the others datasets, C4.5 NP and the C4.5 present equivalent and the best results for Dataset 4, meanwhile RIPPER and NNge have the best result for Dataset 6. MOPSO-N obtained the best average value considering all datasets. The results show that the proposed algorithm is very competitive with other known algorithms from the literature and so, the proposed algorithm has good classification results when dealing with numerical datasets. This conclusion confirms the initial hypothesis: a good classifier can be generated using as objectives the sensitivity and specificity criterion, and MOPSO with numerical data produces good AUC values.

2.6.2 Pareto Front Analysis

The second study was the analysis of the Pareto Front for MOPSO algorithms. This analysis measures which algorithm generates the best set of rules, accordingly to the chosen objectives. Here, the Pareto Front generated by MOPSO algorithms were compared with the real Pareto Front for the chosen dataset.

The real Pareto Front is obtained from the dataset by generating all possible rules, and then, the Pareto Front is built using the sensitivity and specificity. For obvious reasons, this analysis was limited to a small number of datasets (Datasets 2, 5 and 7) and a previous discretization step is needed. The discretization step was performed using a filter from Weka [24]. This filter discretizes a range of numerical attributes in the dataset into discrete attributes by simple binning, that is, the filter divides all ranges in equal parts. The datasets were discretized with three different bins: three, five and seven bins. After the pre-processing step, the real Pareto Front for each dataset discretization were compared with the generated fronts from MOPSO-D. After the comparison of the fronts in the discrete domain, the fronts obtained with MOPSO-N were analyzed. This was made by comparing the numerical front with the three discrete fronts obtained by MOPSO and the "all rules" algorithm.

The rules of all algorithms were obtained considering all the examples of each dataset. Both MOPSO algorithms were executed thirty times and the generated fronts were compared. The comparison was made for the positive and negative rules. Fig. 2.4 shows the Pareto Front for the dataset *ecoli*, for positive rules. This figure presents the fronts of the "all rules" and MOPSO-D for the discretization with three bins and the front of the numerical version of MOPSO. Note that MOPSO-D reached the real Pareto Front and MOPSO-N obtained a better front than the discrete version.

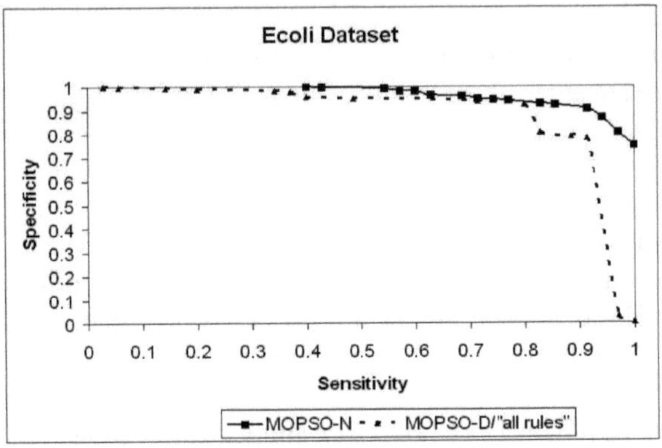

Fig. 2.4. Approximation sets for the `ecoli` dataset, positive rules

To define quantitative measures for the algorithms, the methodology discussed in Section 2.3.1 was adopted. First, the algorithms were compared through the dominance rank [6] using PISA framework [7]. This technique generates rankings and determines the dominance relationship between the algorithm fronts. The statistical test of Mann-Withney [6], U-Test, was applied to verify the significance difference between the ranks. The significance level adopted was 5%. Comparing "all rules" and MOPSO-D, the results showed that for all tests, for all datasets, there is no best Pareto Front. The same is true for the comparison of MOPSO-N and MOPSO-D, and comparing MOPSO-N and "all rules".

Other comparisons were made, now using three indicators: the binary epsilon indicator, hypervolume indicator and R2 indicator [59]. Again, all indicators were executed trough the PISA framework. The reference sets for those indicators were composed by all the non-dominated rules, for all runs. The Mann-Whitney test was used to verify the significance difference for the values of the indicators.

For Dataset 5, the small dataset in analysis with only four attributes, in the comparison of "all rules" and MOPSO-D, there is no difference between the generated fronts. For all discretizations, the indicators do not show a significant difference between the fronts, for positive and negative ones. Comparing MOPSO-N with MOPSO-D, all indicators showed that the MOPSO-D fronts are better than the discrete ones, accordingly U-Test. These results allow us to stress that the MOPSO-D algorithm obtains very good fronts, equivalent to the real Pareto Front. Furthermore, MOPSO-N generates fronts with a better result than "all rules".

The results for Dataset 7 are equivalent to that ones of previous analysis. There is no significant difference for both classes between the fronts generated by "all rules" and MOPSO-D considering the indicators. Again, when comparing only MOPSO-N and MOPSO-D, and MOPSO-N and "all rules", MOPSO-N obtained a better result when comparing the fronts for all indicators.

The last analysis was made with Dataset 2. This was the larger dataset in this comparison, with eight attributes. Here, in the first comparison, MOPSO-D did not obtain equal results to "all rules" for all indicators. Only the epsilon indicator showed an equal result for both algorithms, for both classes. The R2 indicator showed that MOPSO-D obtained a worst front for the dataset with 5 bins, for positive rules, and for the dataset with 3 bins, for negative rules. In the Hypervolume analysis, MOPSO-D had a worst performance for all discretization, for the positive fronts and for the negative ones, it had the worst result for the dataset with three bins. However, comparing MOPSO-N and "all rules" for all discretizations, the first one obtained a better result for all indicators. Furthermore, the approximation set of MOPSO-N overcomes the approximation set of MOPSO-D, for all indicators. This better result of MOPSO-N is presented in Fig. 2.4.

2.7 Predicting Faults

This section presents an application of MOPSO in the Software Engineering domain, specifically in the context of software testing. Software test is an activity related to software quality assurance that has the goal of revealing faults in the program. Testing usually consumes a lot of effort and is a very expensive task. The number of inputs or paths in the program is usually infinite, and the number of classes and methods can be very large. Because of this, strategies to reduce the number of testing goals are fundamental to reduce test costs.

A strategy that has been used is the selection of some classes and methods to be individually tested since to test all classes is in general very expensive in practice. In the literature, there are some works that explores the characteristics of a classes and its fault-proneness [1, 4, 10, 11, 53, 54, 55]. Some of them explore machine learning techniques [20, 22, 39, 42, 55, 58]. A problem with these works is that the obtained models are not easily interpreted for most software developers and users. Most used machine learning techniques can not deal with unbalanced data, as fault datasets are.

In this section, experimental results are presented from an approach based on MOPSO to determine the fault-proneness of a class. This approach was first described in [16]. This experimental study was conducted with the metrics of Table 2.4. They are a widely used sub-set of the Chidamber and Kemerer (CK) metrics suite [12]. The dataset is a free access dataset from NASA [45]. Besides this dataset, we describe next the evaluation measures and the analysis of the obtained results.

Table 2.4. Design Metrics

Metrics	
Coupling Between Objects	CBO
Depth in Inheritance Tree	DIT
Lack of Cohesion of Methods	LCOM
Number of Children	NOC
Response for Class	RFC
Weighted Methods per Class	WMC

Table 2.5. Metrics Description

Metrics Descriptions	
Metric	**Description**
CBO	The number of distinct non-inheritance classes on which a given class is coupled.
DIT	The length of the longest path from a given class to the root in the inheritance hierarchy.
LCOM	The percent of methods of the class which uses the attribute, for each field in class.
NOC	The number of classes that inherit form an specific class.
RFC	The number of methods that could be executed in a given class
WMC	The number of methods implemented with in a give class.

2.7.1 Dataset

The study uses a free-access dataset, available through the NASA Metrics Data Project [45]. The dataset is related to a C++ project (KC1 project). From this dataset, a working dataset was built considering only code implementation faults, leaving out that ones such as configuration or SO errors. Each example of the working dataset is composed by the chosen subset of CK metrics and the goal attribute. The goal attribute is set to true if there is a fault in the class, otherwise is set to false. One aspect of the dataset that has to been highlighted is the unbalanced class distribution. In this dataset 222 instances have the class value true and 1814 has the class value false.

2.7.2 Evaluation Measures

Two analysis were performed. In the first one the classification results of the MOPSO predictor through the Area Under the Curve (AUC) were compared. The second study is the analysis of the metric influence in the fault-proneness prediction of the classes. In this second analysis all rules generated by the classifier and a set of four measures obtained from the training step, presented in Section 2.2, were considered.

2.7.3 Algorithms

Using AUC, MOPSO-N and other four algorithms implemented by Weka tool [24] were compared: C4.5, C4.5 NP, NNge and RIPPER. The experiments were executed using the same methodology presented in Section 2.6: 10-fold-stratified cross validation, for all algorithms the same training and test files were provided and all the attributes have the original numerical values. MOPSO-N was executed with 100 generations and 500 particles for each class. For each fold 30 executions were made and the parameters were set with the same previous values.

2.7.4 AUC Comparison

The AUC values of each algorithm for the KC1 dataset are shown in Table 2.6 and the number between brackets indicates the standard deviation. The cells highlighted present statistical better results. Again, the comparison was made using the non-parametric Wilcoxon test and, the same procedure described before was executed.

Table 2.6. AUC values for KC1 dataset

Algorithm	AUC
MOPSO-N	**73.59 (3.90)**
C4.5	61.13 (10.03)
C4.5 NP	**74.17 (4.87)**
NNge	60.19 (7.18)
RIPPER	57.45 (5.91)

Through the results presented in Table 2.6, it can be observed that MOPSO-N algorithm has a good performance in classification of fault-proneness. MOPSO-N obtains the best AUC values, equivalent to the C4.5 NP algorithm. They present classifications results considerably better than the other algorithms including the random one. C4.5 and NNge classification results are better than a random classifier and could produce good predictions. RIPPER presented the worst result, slightly better than a random classifier. These results allow us to stress that the MOPSO approach can obtain good results in unbalanced dataset and can perform valid predictions.

2.7.5 Influence of C-K Suite Metrics in Fault-Proneness

In the context of fault-prediction the analysis of the rules generated by MOPSO-N is very important. It helps the programmers to establish hypothesis for the influence of each CK metric in the fault-proneness of a class. As said before, the rules are simple and intuitive.

A programmer will be mainly interested in the correct detection of faulty classes. Once, the reputation of the software company can be damaged as it provided a faulty system. Thus, the costs associated with the incorrect classification of a faulty class is clearly higher than the costs associated with the incorrect classification of a correct one. Consequently, for faulty prediction a rule having, a high specificity is more important than a high sensitivity. However, this does not mean that sensitivity can be discarded. A classification technique that classifies all software modules as erroneous might well result in high quality software, the testing costs however will be unjustifiably high. Thus, the analysis of the rules must consider a balance between a high specificity and a reasonable sensitivity. This knowledge can be used by the software manager to efficiently allocate testing resources to fault-prone software modules and increase the quality of the delivered software modules.

The induced rules obtained by MOPSO-N allow to easily make the analysis following the criteria explained below, since, sensitivity and specificity are the objectives used by the algorithm. This analysis used the measures presented in Section 2.2 to define the significance of each metric. If a metric is part of a rule that has good measure values is said that this rule is significant. Furthermore the best rules were selected accordingly their measures. Rules that had a low value of sensitivity and specificity, bellow to a given threshold, were discarded. These thresholds were defined empirically and was set to 0.3 for sensitivity and to 0.5 for specificity. It was defined because its preferable to have rules that covers a few examples than rules that miss to much examples. It was analyzed a total of fifty-three rules obtained through one execution of MOPSO and the best selected rules that gave a larger amount of information are presented in Table 2.7.

Table 2.7. Best Learned Rules Through MOPSO-N

	Rule						
#	**CBO**	**DIT**	**LCOM**	**NOC**	**RFC**	**WMC**	**Class**
1	?	?	$66<LCOM<98$	?	$27 < RFC < 157$	?	faulty
2	?	?	?	?	$37<RFC<152$	?	faulty
3	?	?	?	$NOC\leq 2$	?	$21 < WMC < 99$	faulty
4	$6 < CBO \leq 24$	?	?	?	$35 < RFC < 134$	?	faulty
5	?	?	$89<LCOM<96$	?	$9 < RFC \leq 222$	?	faulty
6	?	?	?	?	?	$27 < WMC < 99$	faulty
7	$8 < CBO \leq 24$	?	?	?	?	$28 < WMC < 99$	faulty
8	?	?	?	$NOC<1$	$65 < RFC < 176$	?	faulty
9	?	?	$93<LCOM\leq 100$	?	$73 < RFC \leq 222$	?	faulty
10	?	?	?	?	$72<RFC<99$	?	faulty

	Measures			
#	**Sens**	**Spec**	**Conf**	**Sup**
1	0.700	0.541	0.172	0.083
2	0.672	0.561	0.172	0.080
3	0.668	0.573	0.175	0.080
4	0.635	0.610	0.181	0.076
5	0.603	0.658	0.194	0.072
6	0.574	0.717	0.216	0.068
7	0.514	0.778	0.240	0.061
8	0.481	0.803	0.250	0.057
9	0.417	0.869	0.303	0.049
10	0.331	0.911	0.338	0.03

Rules 2, 6 and 10 of Table 2.7 show that RFC and WMC are the most important to indicate a fault in the class. Values for these metrics in the intervals stated in the rules imply in a high probability of fault in the class. These rules are the only ones with just one attribute, with good sensitivity values and high specificity values. These last ones says that the rules do not cover negative examples.

Observing the rules 1, 4, 5, 7 and 9 we can conclude that great values for CBO and LCOM, associated to great values of RFC or WMC, are related to the presence of a fault in the class. Considering NOC, we can note that there two rules, 3 and 8, also associated to RFC and WMC metrics. This rule is more specific, and can cover more examples with a less number of errors. There is no rule related to the metric DIT, we can conclude that its value is not significant for fault-proneness.

2.8 Related Works

Several works were proposed to find a set of rules for the classification task. This section describes two lines of works more related to our paper: the subselection algorithms and the multiobjective meta-heuristic works.

The works [5], [37], [38],[44], and [57] use association rule algorithms that usually generates a huge number of rules. Due to this in [32] an algorithm, named AprioriC is presented to remove redundant rules. However, this algorithm continues generating a large set of rules. In [44] another algorithm named ROCCER is introduced. It constructs a convex hull in ROC space and uses the Apriori algorithm with Minimum support and confidence to generate a large set of rules. The idea is to insert a rule that leads to a point outside the current ROC convex hull. In addition to the need of the Apriori algorithm, the insertion and the removal of rules imply on backtracking, a time consuming procedure for the ROCCER algorithm. Using the set of rules as the classification model, the Pareto Front Elite algorithm [30] has the same goal of the algorithms detailed above, i.e., to maximize the AUC. One association rule algorithm is executed to generate rules with support greater than a minimum support parameter or confidence greater than a minimum confidence parameter. From the rule set, the algorithm uses the sensitivity and the specificity criteria in a post-process method to select the Pareto Front. If the minimum parameters are set to very low values then this algorithm can be considered as an all rules approach, i.e., all the rules with high sensitivity or specificity values are in the classifier induced by the Pareto Front Elite algorithm.

Recently, increasing interest has emerged in applying the concept of Pareto-optimality to machine learning inspired by the successful developments in evolutionary multiobjective optimization. These researches include multiobjective feature selection, multiobjective model selection in training multilayer perceptrons, radial-basis-function networks, support vector machines, decision trees and intelligent systems [31]. In the literature, few works deal with multiobjective evolutionary algorithms for rule learning [28], [18] and [27]. The first work focuses on the rule selection phase. It presents a genetic-based multiobjective rule selection algorithm to find a smaller rule subset with higher accuracy than the rule sets extracted with heuristic approaches. The algorithm has the objective to maximize the accuracy, and to minimize the number of rules. In [27], multiobjective association rules generation and selection with NSGA-II (Non-Dominated Sorting Genetic Algorithm) are discussed. In [17], a multiobjective optimization evolutionary algorithm with Pareto concepts is used to discover classification rules for a target class. It presents an implementation of NSGA with positive confidence and sensitivity as objectives. This work is extended in [18] by using Multi-Objective Meta heuristics to produce sets of interesting classification rules. A measure of dissimilarity of rules was introduced to promote diversity on the population.

In the other hand, this work uses Multi-Objective Meta heuristics to conceive a complete novel approach to induce classifiers, where the properties of the rules can be expressed in different objectives, and then the algorithm finds these rules in an unique run by exploring Pareto dominance concepts. Furthermore, these rules can be used as an unordered classifier, in this way, the rules are more intuitive and easier to understand because they can be interpreted independently. The quality of the learned rules is not affected during the learning process because the dataset is not modified, as in traditional rule induction approaches. With this philosophy, this work describes a Multi-Objective Particle Swarm Optimization (MOPSO) algorithm. One reason to choose the Particle Swarm Optimization Meta heuristic is its recognized ability to work in numerical domains. This propriety allows the described algorithm to deal with both numerical and discrete attributes.

2.9 Conclusions

In this work, a rule learning algorithm based on multi-objective particle swarm optimization is presented. Three main challenges were addressed. First, the use of Multi-Objective approach allows us to create classifiers in only one step. This method works finding the best non-dominated rules of a problem, by selecting them in the rules generation process. It is a simplified way to obtain more intelligible rules from a database and gives certain freedom to choose new objectives like interest or novelty. Second, the algorithm deals with both numerical and discrete data, and third, the induced classifiers present good results in terms of the AUC metric.

The approach was empirical evaluated aiming to confirm two different features of the algorithm. The first one is related to the induced classifiers, and compares MOPSO-N with four well know algorithms from the literature: C4.5, C4.5 No Pruning, RIPPER and NNge. The results show that the Multi-Objective Particle Swarm Rule Learning is very competitive with respect to the other algorithms. Furthermore, once the chosen objectives were the sensitivity and the specificity criteria, it is possible to affirm that rules with these properties create efficient classifiers in terms of the AUC metric. In the second analysis we compared the approximation set of both MOPSO algorithms. We used the dominance rank and three indicators: epsilon, hypervolume and R. The results show a better performance of MOPSO-N.

Finally, an application in prediction fault was presented. The use of MOPSO-N allows to analyze the influence of each design metric in fault-prediction. The results point out that great values of RFC and WMC can imply a high probability of fault in the class. Thus, the rules produced are very useful and can be directly interpreted. They can be used as a strategy for reducing effort testing and focusing fault-proneness classes.

Future works include: the execution of a greater number of experiments to validate the initial results, the comparison of our approach with other known algorithms in the literature, and enhancements of the algorithm to profit a more diverse set of rules without increasing the size of the rule set. The algorithm proposed also has a lot of possible applications for diverse problems in the Software Engineering area, and they should be explored.

Acknowledgment

This work was supported by the CNPQ (project: 471119/2007-5).

References

1. Alshayeb, M., Li, W.: An empirical validation of object-oriented metrics in two different iterative software processes. IEEE Transaction on Software Engineering 29(11), 1043–1049 (2003)
2. Asuncion, A., Newman, D.: UCI machine learning repository (2007)
3. Baronti, F., Starita, A.: Hypothesis Testing with Classifier Systems for Rule-Based Risk Prediction, pp. 24–34. Springer, Heidelberg (2007), `http://dx.doi.org/10.1007/978-3-540-71783-6_3`
4. Basili, V.R., Briand, L.C., Melo, W.L.: A validation of object-oriented design metrics as quality indicators. IEEE Transaction on Software Engineering 22(10), 751–761 (1996)
5. Batista, G., Milare, C., Prati, R.C., Monard, M.: A comparison of methods for rule subset selection applied to associative classification. Inteligencia Artificial. Revista Iberoamericana de IA 7(32), 29–35 (2006)
6. Bleuler, S., Laumanns, M., Thiele, L., Zitzler, E.: PISA — a platform and programming language independent interface for search algorithms. In: Fonseca, C.M., Fleming, P.J., Zitzler, E., Deb, K., Thiele, L. (eds.) EMO 2003. LNCS, vol. 2632, pp. 494–508. Springer, Heidelberg (2003)
7. Bleuler, S., Laumanns, M., Thiele, L., Zitzler, E.: The PISA homepage (2003), `http://www.tik.ee.ethz.ch/pisa/`
8. Bradley, A.P.: The use of the area under the ROC curve in the evaluation of machine learning algorithms. Pattern Recognition 30(7), 1145–1159 (1997)

9. Bratton, D., Kennedy, J.: Defining a standard for particle swarm optimization. In: Proceedings of IEEE Swarm Intelligence Symposium (SIS 2007), Honolulu, Hawaii, USA, pp. 120–127. IEEE Computer Society, Los Alamitos (2007)
10. Briand, L.C., Wust, J., Daly, J., Porter, V.: A comprehensive empirical validation of design measures for object-oriented systems. In: METRICS 1998: Proceedings of the 5th International Symposium on Software Metrics, Washington, DC, USA, p. 246. IEEE Computer Society, Los Alamitos (1998)
11. Briand, L.C., Wust, J., Daly, J.W., Porter, D.V.: Exploring the relationships between design measures and software quality in object-oriented systems. The Journal of Systems and Software 51(3), 245–273 (2000)
12. Chidamber, S., Kemerer, C.: A metrics suite for object-oriented design. IEEE Transaction on Software Engineering 20(6), 476–493 (1994)
13. Clark, P., Niblett, T.: Rule induction with CN2: Some recent improvements. In: ECML: European Conference on Machine Learning. Springer, Heidelberg (1991)
14. Cohen, W.W.: Fast effective rule induction. In: Proceedings of the Twelfth International Conference on Machine Learning, pp. 115–123 (1995)
15. Conover, W.J.: Practical nonparametric statistics. Wiley, Chichester (1971)
16. de Carvalho, A.B., Pozo, A., Vergilio, S., Lenz, A.: Predicting fault proneness of classes trough a multiobjective particle swarm optimization algorithm. In: Poceedings of 20th IEEE International Conference on Tools with Artificial Intelligence (2008)
17. de la Iglesia, B., Philpott, M.S., Bagnall, A.J., Rayward-Smith, V.J.: Data mining rules using multi-objective evolutionary algorithms. In: Congress on Evolutionary Computation, pp. 1552–1559. IEEE Computer Society, Los Alamitos (2003)
18. de la Iglesia, B., Reynolds, A., Rayward-Smith, V.J.: Developments on a multiobjective metaheuristic (momh) algorithm for finding interesting sets of classification rules. In: Coello, C.A.C., Hernández Aguirre, A., Zitzler, E. (eds.) EMO 2005, vol. 3410, pp. 826–840. Springer, Heidelberg (2005)
19. Egan, J.: Signal detection theory and ROC analysis. Academic Press, New York (1975)
20. Pérez-Miñana, E., Gras, J.-J.: Improving fault prediction using bayesian networks for the development of embedded software applications: Research articles. Softw. Test. Verif. Reliab. 16(3), 157–174 (2006)
21. Fawcett, T.: Using rule sets to maximize ROC performance. In: IEEE International Conference on Data Mining, pp. 131–138. IEEE Computer Society Press, Los Alamitos (2001)
22. Fenton, N., Neil, M., Marsh, W., Hearty, P., Marquez, D., Krause, P., Mishra, R.: Predicting software defects in varying development lifecycles using bayesian nets. Infromation on Software Technology 49(1), 32–43 (2007)
23. Ferri, C., Flach, P., Hernandez-Orallo, J.: Learning decision trees using the area under the ROC curve. In: Sammut, C., Hoffmann, A. (eds.) Proceedings of the 19th International Conference on Machine Learning, July 2002, pp. 139–146. Morgan Kaufmann, San Francisco (2002)
24. Group, W.M.L.: Weka machine learning project (2007), `http://www.cs.waikato.ac.nz/ml/weka`
25. Hanley, J.A., McNeil, B.J.: The meaning and use of the area under a receiver operating characteristic (ROC) curve. Radiology 143(1), 29–36 (1982)
26. Hansen, M.P., Jaszkiewicz, A.: Evaluating the quality of approximations to the non-dominated set. Technical Report IMM-REP-1998-7, Technical University of Denmark (March 1998)

27. Ishibuchi, H.: Multiobjective association rule mining. In: PPSN Workshop on Multiobjective Problem Solving from Nature, Reykjavik, Iceland, pp. 39–48 (2006)
28. Ishibuchi, H., Nojima, Y.: Accuracy-complexity tradeoff analysis by multiobjective rule selection. In: ICDM, pp. 39–48. IEEE Computer Society, Los Alamitos (2005)
29. Ishida, C., de Carvalho, A.B., Pozo, A.: Exploring Multi-objective PSO and GRASP-PR for rule induction. In: van Hemert, J., Cotta, C. (eds.) EvoCOP 2008. LNCS, vol. 4972, pp. 73–84. Springer, Heidelberg (2008)
30. Ishida, C.Y., Pozo, A.T.R.: Optimization of the auc criterion for rule subset selection. In: 7th. International Conference on Intelligent Systems Design and Applications, New York, NY, USA. IEEE Computer Society, Los Alamitos (2007)
31. Jin, Y.: Multi-Objective Machine Learning. Springer, Berlin (2006)
32. Jovanoski, V., Lavrac, N.: Classification rule learning with APRIORI-C. In: Brazdil, P.B., Jorge, A.M. (eds.) EPIA 2001. LNCS, vol. 2258, pp. 44–51. Springer, Heidelberg (2001)
33. Kennedy, J., Eberhart, R.: Particle swarm optimization. In: IEEE International Conference on Neural Networks, pp. 1492–1948. IEEE Press, Los Alamitos (1955)
34. Kennedy, J., Eberhart, R.C.: Swarm intelligence. Morgan Kaufmann Publishers Inc., San Francisco (2001)
35. Knowles, J., Thiele, L., Zitzler, E.: A Tutorial on the Performance Assessment of Stochastic Multiobjective Optimizers. In: Computer Engineering and Networks Laboratory (TIK), ETH Zurich, Switzerland, Febuary 2006, vol. 214 (2006) (revised version)
36. Lavrac, N., Flach, P., Zupan, B.: Rule evaluation measures: A unifying view. In: Džeroski, S., Flach, P.A. (eds.) ILP 1999. LNCS, vol. 1634, pp. 174–185. Springer, Heidelberg (1999)
37. Li, W., Han, J., Pei, J.: CMAR: Accurate and efficient classification based on multiple class-association rules. In: Cercone, N., Lin, T.Y., Wu, X. (eds.) ICDM, pp. 369–376. IEEE Computer Society, Los Alamitos (2001)
38. Liu, B., Hsu, W., Ma, Y.: Integrating classification and association rule mining. In: Knowledge Discovery and Data Mining, pp. 80–86 (1998)
39. Lounis, H., Ait-Mehedine, L.: Machine-learning techniques for software product quality assessment. In: Fourth International Conference QSIC 2004: Proceedings of the Quality Software, Washington, DC, USA, pp. 102–109. IEEE Computer Society, Los Alamitos (2004)
40. Martin, B.: Instance-Based learning: Nearest Neighbor With Generalization. PhD thesis, Department of Computer Science, University of Waikato, New Zealand (1995)
41. Mostaghim, S., Teich, J.: Strategies for finding good local guides in multi-objective particle swarm optimization. In: Proceedings of the 2003 IEEE Swarm Intelligence Symposium SIS 2003 Swarm Intelligence Symposium, pp. 26–33. IEEE Computer Society, Los Alamitos (2003)
42. Pai, G.J., Dugan, J.B.: Empirical analysis of software fault content and fault proneness using bayesian methods. IEEE Transaction on Software Engineering 33(10), 675–686 (2007)
43. Pareto, V.: Manuel d"economie politique (1927)

44. Prati, R.C., Flach, P.A.: ROCCER: An algorithm for rule learning based on ROC analysis. In: Proceedings of the Nineteenth International Joint Conference on Artificial Intelligence, pp. 823–828 (2005)
45. Program, N.I.F.M.D.: Metrics data repository, `http://mdp.ivv.nasa.gov/`
46. Provost, F., Fawcett, T.: Robust classification for imprecise environments. Machine Learning 42(3), 203 (2001)
47. Provost, F., Fawcett, T., Kohavi, R.: The case against accuracy estimation for comparing induction algorithms. In: Proceedings 15th International Conference on Machine Learning, pp. 445–453. Morgan Kaufmann, San Francisco (1998)
48. Provost, F.J., Fawcett, T.: Analysis and visualization of classifier performance: Comparison under imprecise class and cost distributions. In: KDD, pp. 43–48 (1997)
49. Quinlan, R.: C4.5: Programs for Machine Learning. Morgan Kaufmann Publishers, San Mateo (1993)
50. Rakotomamonjy, A.: Optimizing area under roc curve with SVMs. In: Hernández-Orallo, J., Ferri, C., Lachiche, N., Flach, P.A. (eds.) ROCAI, pp. 71–80 (2004)
51. Reyes-Sierra, M., Coello, C.A.C.: Multi-objective particle swarm optimizers: A survey of'the state-of-the-art. International Journal of Computational Intelligence Research 2(3), 287–308 (2006)
52. Sebag, M., Aze, J., Lucas, N.: ROC-based evolutionary learning: Application to medical data mining. In: International Conference on Artificial Evolution, Evolution Artificielle. LNCS, vol. 6 (2003)
53. Subramanyam, R., Krishnan, M.S.: Empirical analysis of CK metrics for object-oriented design complexity: Implications for software defects. IEEE Transaction on Software Engineering 29(4), 297–310 (2003)
54. Succi, G., Pedrycz, W., Stefanovic, M., Miller, J.: Practical assessment of the models for identification of defect-prone classes in object-oriented commercial systems using design metrics. The Journal of Systems and Software 65(1), 1–12 (2003)
55. Thwin, M.M.T., Quah, T.-S.: Application of neural networks for software quality prediction using object-oriented metrics. The Journal of Systems and Software 76(2), 147–156 (2005)
56. Toracio, A., Pozo, A.: Multiple objective particle swarm for classification-rule discovery. In: Proceedings of CEC 2007, pp. 684–691. IEEE Computer Society, Los Alamitos (2007)
57. Yin, X., Han, J.: CPAR: Classification based on predictive association rules. In: Proceedings SIM International Conference on Data Mining (SDM 2003), pp. 331–335 (2003)
58. Zhou, Y., Leung, H.: Empirical analysis of object-oriented design metrics for predicting high and low severity faults. IEEE Transaction on Software Engineering 32(10), 771–789 (2006)
59. Zitzler, E., Thiele, L.: Multiobjective Evolutionary Algorithms: A Comparative Case Study and the Strength Pareto Approach. IEEE Transactions on Evolutionary Computation 3(4), 257–271 (1999)
60. Zitzler, E., Thiele, L., Laumanns, M., Fonseca, C.M., da Fonseca., V.G.: Performance assessment of multiobjective optimizers: an analysis and review. IEEE Transactions on Evolutionary Computation 7, 117–132 (2003)

3

Use of Multiobjective Evolutionary Algorithms in Water Resources Engineering

Francisco Venícius Fernandes Barros,
Eduardo Sávio Passos Rodrigues Martins,
Luiz Sérgio Vasconcelos Nascimento, and Dirceu Silveira Reis Jr.

Research Institute for Meteorology and Water Resources
FUNCEME, Av. Rui Barbosa, 1246, Fortaleza, CE, CEP: 60115-221, Brazil
{veniciusfb,espr.martins,luizsergiovn,dirceu.reis}@gmail.com

In Engineering, and more specifically in water resources, the need of representation of complex natural phenomena through models is of crucial importance for water resources planning and management. Through the use of these models, it is possible to understand the natural processes and to evaluate the system response to different scenarios, providing support to the decision making process. In this chapter, we investigate the use of this models to water resource engineering.

3.1 Introduction

Among existing models, two classes are of great importance to water resources planning and management: the rainfall-runoff models and the reservoir system operation models. The latter class of models makes use of either systematic records of reservoir's inflows or simulated series obtained by the first class of models, the hydrologic models, which are mathematical representations of the natural processes that occur in a watershed.

With respect to rainfall-runoff models, in order to represent satisfactorily the natural process, their parameters should be determined appropriately. In most cases, these parameters cannot be directly determined due to either the impossibility of thei estimation in the field or their abstract nature. The indirect determination of the parameters can be made through a calibration study of the model under analysis as long as one has a common period of series that represents both the input and output of this model. In a calibration study of a rainfall-runoff model, the parameters are chosen such that the simulated streamflow is as close as possible to the observed streamflow. It

N. Nedjah et al. (Eds.): Multi-Objective Swarm Intelligent Systems, SCI 261, pp. 45–82.
springerlink.com

is often the case that an objective metric is used to measure the degree of closeness between these two series, providing the basis for comparison among different sets of model parameters.

Calibration can be carried out either manually or automatically. The manual method consists of a series of trial-and-error attempts, in which the parameters are chosen based on the hydrologist experience and knowledge of the region of study. After this choice, the hydrologic model is run and then a comparison is made between the observed and the simulated hydrographs, subjectively, looking after the set of parameters that produces the best result. Solutions of such nature are generally very demanding in terms of work and time, besides requiring the full knowledge of those models, which sometimes are extremely complex.

The automatic calibration is based upon the use of optimization algorithms that perform a search for the optimal solution with respect to one or more objectives. Several research studies were done in the last decades in this scope, and experience has shown that optimal search based on only one objective, regardless of how carefully it can be made, are not able to determine a solution that models adequately the phenomenon under study. Another factor that favors a multiobjective optimization is that real-world problems frequently require the analysis of conflicting multiobjectives. For such cases, it is possible to use the Pareto front concept, which makes possible the comparison of solutions with multiple objectives.

The other class of models that is extremely important for water resources engineering is the reservoir system operation models. This type of models is also used here to explore the potential of using multiobjective optimization algorithms to determine optimal operation policies with respect to prespecified objectives. The definition of how to operate a system with several reservoirs is a complex task because it includes many technical, social and political aspects, and involves multiple inter-related decisions in time ([1],[2]). One important aspect of this complexity is the existence of multiple objectives, usually conflicting ones, such as meeting water supply and irrigation demands, energy generation, maintenance of aquatic species, flood control, navigation, etc.

Defining reservoirs' operating rules usually means to specify the volume of water that should be released by each reservoir over time. These rules are often specified in order to maximize or minimize one or more objective functions that translate the goals of operating a reservoirs' system. This optimization study has to respect some constraints, such as reservoir storage capacity, maximum pumping rate, maximum flow rate, minimum flow in river, etc. As the system gets more complex, it becomes very hard to find an optimal operating policy for the system.

Among the algorithms used nowadays, a group in particular has been topic of several research studies due mainly to its widespread use in several areas of science, commerce and engineering, and also due to the easiness of their implementation: the evolutionary algorithms. A great advantage of this class of algorithms relative to other approaches is the fact they work simultaneously with a set of solutions, allowing a global perspective of the problem, a greater diversity in the search, and more reliable solutions. Moreover, evolutionary algorithms do not depend on specific characteristics of the objective function to properly work, such as, concavity, convexity and continuity. Besides, these algorithms allow a more comprehensive investigation of the parameter space at each iteration, reducing the chances of being trapped in a local minimum (maximum). Their intrinsic stochastic nature provides a more diverse population of possible solutions, allowing the construction of a Pareto front in a single run of the algorithm, which makes them a good option for multiobjective problems.

3.2 Literature Review

The identification of the optimum of uni-modal functions is a problem for which several strategies have been explored in the literature. However, in practice, uni-modal function is barely the case, since most of real-world problems have several local solutions, while only one is the global optimum. Within a context of calibration of hydrologic models, an additional challenge emmerges once one is interested in identifying a set of parameters of a hydrologic model that better represents the behavior of streamflow generation process in a watershed through time ([3]). Moreover, the involved subjectivity and the required time for manually fit a hydrologic model on a trial-error basis motivated the intense research on automatic calibration of hydrologic models ([4]).

However, most studies focused on the uni-objective automatic hydrologic model calibration ([5],[6],[7]), while the multiobjective approach has been explored only in the last decade. Within a multiobjective context, a great variety of evolutionary algorithms has been developed based on the concurrent evolution of multiple non-dominate solutions at each iteration of these algorithms ([8]).

Different multiobjective approaches have been employed in the implementation of evolutionary algorithms ([9]), such as: objective weighting, lexicographic ordination, use of sub-populations (each one being a uni-objective optimizer), Pareto concept and possible combinations of the previous alternatives. The implementation of multiobjective approaches in evolutionary algorithms has followed this taxonomy, as shown by: 1. Objective weighting ([10], [11], [12]); 2. Lexicographic ordination ([13]); 3. Use of sub-populations ([14]); 4. Pareto concept ([15], [16], [17]); and, 5. Possible combinations of the previous alternatives ([18]).

Within a hydrologic modeling context, two developments can be highlighted: the multiobjective Complex Evolution (MOCOM-UA; [19]) and the multiobjective Shuffled Complex Evolution Metropolis (MOSCEM-UA; [4]).

The MOCOM-UA resolves the multiobjective calibration problem by the application of the Pareto concept to the global Shuffled Complex Evolution algorithm (SCE-UA; [5], [6]). MOSCEM was developed in order to better identify the Pareto front, specially its extremes, as well as to resolve the MOCOM-UA deficiency regarding the premature convergence in the presence of a large number of parameters and highly correlated objectives ([19], [20], [21], [22], [23], [24]).

With regard to optimization of reservoir operation, in the last decade, there has been observed a great effort of the community to develop global search algorithms. Evolutionary algorithms has been applied quite successfully in many engineering problems ([25], [26], [27], [28], [29]).

In this chapter, two evolutionary algorithms are used for the uni-objective case, the Honey Bee Mating Optimization (HBMO) and the Particle Swarm Optimization (PSO), while three algorithms are employed for the multiobjective case, the multiobjective Honey Bee Mating Optimization (MOHBMO), the multiobjective Particle Swarm Optimization (MOPSO) and the multiobjective Shuffled Complex Evolution Metropolis (MOSCEM). The HBMO and MOHBMO are based upon the honey bee mating flights, PSO and MOPSO algorithms are based on the social behavior of individuals, while the SCEM and the MOSCEM are based upon Markov Chain Monte Carlo methods (MCMC). The uni-objective version HBMO is that proposed by [30], while its multiobjective version MOHBMO is proposed by the present study. First, these algorithms are tested with mathematical functions which represent a challenge for any optimization algorithm, named here simply test functions ([29]).

This study focus on the application of these evolutionary algorithms for calibration of hydrologic models and identification of optimal operating policies for the reservoir's system that supply water for the Metropolitan Region of Fortaleza (MRF) and other smaller local demands.

3.3 Evolutionary Algorithm

Evolutionary algorithms include search methods that have their insight on natural processes, such as: behavior of social groups, animal reproduction, among others. These algorithms are based on the "survival of the fittest", meaning that the best solutions will prevail over the others. These algorithms have characteristics that make them more robust than other approaches to search for optimal solutions, among which can be highlighted:

- The ability to work simultaneously with a population of solutions, which introduces a global perspective and a greater diversity in the search. Such characteristic promotes a great capacity to find a global optimal in problems that have several local optimal solutions;

- Unlike other algorithms based on differential calculus or other specific procedure, evolutionary algorithms work with any objective function and require no specific characteristic, such as continuity, concavity or convexity;
- No previous knowledge of the search space is needed. The search space can be multidimensional, constrained or not, either linear or not.

3.3.1 Uni-objective Optimization

Honey Bee Mating-based Optimization (HBMO)

The uni-objective version of the algorithm (HBMO) is the one proposed by [30], in which the honey bee mating has served as inspiration. The relationship between the natural process and the algorithm is established as the algorithm is described mathematically in the sequence.

The algorithm starts with an initial population (hive), composed by a set of solutions randomly sampled from a Uniform Distribution. A fitness value is assigned to each solution of the initial population, equals to the selected objective function. The best solution (queen) is then selected based on the fitness value, the smallest fitness value since it is a minimization problem. All other solutions are discarded and a new iteration is initiated. If the problem at hand is a maximization problem, then the following equation should be used:

$$max f(\mathbf{x}) = min - f(\mathbf{x}) \tag{3.1}$$

At the beginning of a new iteration, random solutions (drones, D) with certain degree of dependence with the best solution (queen, Q) are generated. Such dependence is a linear or a quadratic function of the number of iterations (see equations 3.2 and 3.3), increasing with the number of iterations (maturity of the hive). At the last iteration there is a large dependence among the drones and the queen, which promotes the convergence of the search. In order to guarantee the diversity of the solutions, a lower limit for the term δ^2/nMF^2 was introduced. Here the lower limit was set to 0.1.

$$D = Q \times [1 - \delta/nMF] + d \times [\delta/nMF] \tag{3.2}$$

$$D = Q \times [1 - (\delta^2/nMF^2)] + d \times (\delta^2/nMF^2) \tag{3.3}$$

where nMF is the maximum number of iterations, i is the actual iteration, and d is a random solution in the search space. The set of random solutions $\{d\}$ is centered at the best solution (queen), from which each random solution d is generated. The parameter δ is given by the following expression:

$$\delta = nMF - (i - 1) \tag{3.4}$$

At each iteration, a selective test (mating flight) is performed in order to determine probabilistically whether or not the best solution (queen) will receive information (mating) from the randomly selected solutions (drones). This is done by applying an annealing function, also known as Boltzman function, as suggested by [31]:

$$p(Q, D) = exp(-\Delta(f)/S_p^{(t)}) \tag{3.5}$$

where $p(Q, D)$ is the probability that the best solution Q receives information from the selected random solution D (crossover between the drone D and the queen Q), $\Delta(f)$ is the absolute difference of the fitness values of solutions Q and D, and S_p^t is the temperature (speed of the queen) of the annealing function at time t (during flight). Looking at the annealing function, it is evident that the probability is larger when either the temperature (speed of the queen) is high or the differences in fitness are small (the fitness of the drones are close to the fitness of the queen).

The algorithm allows information exchange (mating) between the current solutions (drones and queen) with probability $p(Q, D)$. In case of information exchange (mating), the information of the solution (genetic information of the drone) is selected and stored in a repository (queen's spermatheca), and the temperature of annealing (speed of the queen) decreases as follows:

$$S_p^{(t+1)} = \alpha^{(t)} \times S_p^{(t)} \tag{3.6}$$

and

$$\alpha^{(t)} = [M - m^{(t)}]/M \tag{3.7}$$

where $S_p^{(t)}$ is the temperature (speed of the queen) at time t, $\alpha^{(t)}$ is a value between 0 and 1, M is the size of the repository (queen's spermatheca) and $m^{(t)}$ is the number of randomly selected solutions (drones) for the crossover. The number of information exchange attempts (queen's energy) decreases as follows:

$$E^{(t+1)} = E^{(t)} - \gamma \tag{3.8}$$

where $E^{(t)}$ is the number of attempts at time t and γ its decay at each time interval. For this study, γ is equals to one.

The best solution (queen) is likely to receive information (mating) as long as both the number of attempts (its energy) is not close to zero and its repository (spermatheca) is not full. From equation (3.8), it is noticed that the decay value γ will determine how many tests (transitions in the search space) the best solution (queen) can perform at each selection of random solutions (mating flight). Other limiting factors are the temperature of the annealing function (speed of the queen), which should be greater than zero, and the number of random solutions (drones) available for testing, since each random solution only provides information (mating) once and then is discarded (death of the drone).

The generation of new solutions (offspring) by crossover occurs after the information exchange between the information stored in the repository (genes of drones in the spermatheca) and the information of the best solution (genes of the queen). The choice of which information (genetic material in the spermatheca) will be used is randomly determined and can be reused. This generation process is carried out using several crossover operators according to their performance, here evaluated by the percentage of their contribution to the generation of new solutions. It was used two crossover operators as described below: the Arithmetic Crossover and the Blend Crossover.

Once new solutions (new offspring) are generated, an attempt is made to improve both the new solutions and the best solution by use of a mutation procedure, described below, and known as Creep mutation operator. The mutation is randomly applied to a pre-specified percentage of the new solutions (new offspring). Also, there is a probability that mutation is applied to the best solution (queen), set herein to 5%.

After mutation, the population is then evaluated based on the objective function. If the best generated solution is better than the best current solution (queen), the best solution is updated, otherwise, the best solution continues the same. At each iteration, all generated solutions are discarded and only the best solution is kept.

The process described previously is repeated until a stop criterion is met, such as the maximum number of iterations. As an attempt to perform a more detailed search, [30] suggest the use of several queens, selected based on their fitness values. In such case, the process described earlier is applied for each queen, mixing all offspring afterwards. The best solutions are then selected from the sets formed by both queens and their offspring. This approach results in a refinement over the original approach based on only one queen, therefore, it is used in this study.

Crossover and Mutation Operators: The crossover operators employed here are the Blend Crossover and the Arithmetic Crossover operators. The Blend Crossover performs a linear combination between two solutions as indicated by the following expression ([32]):

$$c = p_1 + \beta \times (p_2 - p_1) \tag{3.9}$$

where c is the generated offspring, p_1 and p_2 are the parent solutions, and β represents the feasible space for offspring generation with $\beta \sim U(-\varepsilon, 1+\varepsilon)$, where a previously chosen ε allows the generation to occur beyond the interval defined by p_1 and p_2.

The Arithmetic Crossover does a linear combination between two solutions according to the following expressions ([32]):

$$c_1 = \beta \times p_1 + (1 - \beta) \times p_2 \tag{3.10}$$

$$c_2 = (1 - \beta) \times p_1 + \beta \times p_2 \tag{3.11}$$

where $\beta \sim U(0, 1)$.

The mutation operator corresponds to the Creep Mutation operator, which performs a small perturbation in one decision variable. This perturbation is carried out according to the following expression:

$$c_n^i = \beta \times c_n^i \tag{3.12}$$

Particle Swarm Optimization (PSO)

MOPSO is a multiobjective version of the uniobjective Particle Swarm Optimization introduced by [33], inspired by the behavior of social groups, such as those of birds, fishes and insects. The version of MOPSO employed here is the one proposed by [34]. This section starts with a brief description of the PSO followed by the modifications that were made to allow the algorithm to deal with multiobjective cases.

Initially, the algorithm randomly generates a set of solutions (particles) within the feasible space. Each solution is assigned the value of the objective function, which is used as a metric of its fitness. Then, the algorithm selects the best solution among those contained in the initial set of solutions. In this case, the best solution (individual) is the one that has the least value of the objective function. The algorithm also considers the best solution as the best global solution (Swarm leader). The algorithm uses the concept of best individual of each solution (particle), which is the best position up to the current iteration in the evolution of the search. In the beginning, the best individual of each solution is its initial value. Each particle (solution) of the swarm of N particles has a current position, at the current iteration, and a given velocity, which is updated according to the particle's and group's experiences. Vector x, which contains the positions of all particles of the population, can be computed at each iteration as follows,

$$x^{(t+1)} = x^{(t)} + \chi\nu^{(t)} + \varepsilon^{(t)} \tag{3.13}$$

where $x^{(t)}$ and $x^{(t+1)}$ are the vectors that contain the positions of the N particles (solutions) at iterations t and $t+1$, respectively, $\nu^{(t)}$ is the vector whose elements represent the velocity of the N particles at time t, χ is a factor that controls the magnitude of the velocities (between 0 and 1), and $\varepsilon^{(t)}$ is a small stochastic perturbation known as "turbulence factor", which helps the algorithm to avoid local optima and to increase the diversity of the search.

The velocity of each particle is updated at each iteration by the combination of two terms: the best position of the particle, contained in the vector P, which explores the best result experienced by the particle up to the current iteration, and the best global position, contained in the vector G, which is the best solution achieved by the whole population up to current iteration. The velocity vector of size [N,1] is computed by the following expression,

$$\nu^{(t+1)} = w\nu^{(t)} + c_1 r_1 (P - x^{(t)}) + c_2 r_2 (G - x^{(t)}) \tag{3.14}$$

where w is the inertia of the particle, c_1 and c_2 are constants that control the influence of the individual and global velocities, and r_1 and r_2 are uniformly generated random numbers between [0, 1]. This study employed the following values of these parameters: $c_1 = c_2 = 1$, and w varying linearly between 0.95 and 0.4 in the first 70% of the maximum number of iterations, and equal to 0.4 for the remaining iterations. The values of these parameters were determined in a sensitivity analysis study.

The algorithm runs until the number of iterations reaches the maximum number of iterations specified by the user.

Shuffled Complex Evolution Metropolis (SCEM)

The Shuffled Complex Evolution Metropolis algorithm (SCEM) is an adapted version of the global optimization algorithm SCE-UA developed by [5], well suited for calibration of hydrologic models.

The SCEM algorithm is based on Bayesian Theory, and as such, treats the parameters of hydrological models as statistical variables, and looks for a posterior probability distribution from the likelihood function and a priori distribution of the same parameters. The algorithm evolves by looking for solutions in the search space which maximizes the likelihood of observing such parameters. The resulting posterior probability distribution captures the probabilistic behavior of the parameters.

Initially, a set of solutions in the search space $(\theta_1, \theta_2, \cdots, \theta_s)$ is generated from a *priori* distribution, and the posterior probability is computed for each point (or the objective functions for the case). The posterior probability computation follows a Bayesian inference scheme proposed by [35].

The s points $(\theta_1, \theta_2, \cdots, \theta_s)$ are then classified in decreasing order according to the posterior probability, so that the first solution will correspond to the posterior probability highest value. These points are stored in an array $D_{(s \times n+1)}$, which represents a set of ranked solutions based upon the posterior probability (n is the number of parameters).

Parallel sequences with same number of solutions are initialized from the set of ranked sequences, $S^k = D_{(k \times n+1)}$ with $k = 1, 2, \cdots, q$, where q is the number of complexes. The s points of D are, then, partitioned into q sets, said to be complexes, $C^q = D_{(q(j-1) \times n)}$ with $j = 1, 2, \cdots, m$, where m is the number of solutions in each complex. The kth complex contains every $q(j-1)+k$ ranked point.

The parallel sequences form the set of solutions used in the search for the best solution, providing the algorithm the ability to independently explore the search space. The use of these independent search groups allow the algorithm to deal with more than one region of attraction and, heuristically, decide if the sequences are converging.

The use of complexes allows grouping information over the search space, which is obtained during the evolution process performed in each sequence of the SEM (Sequence Evolution Metropolis) algorithm. The SEM algorithm is described in detail below.

After the run of the SEM algorithm, the solutions that form the complexes are classified in decreasing order of probability estimated based upon the posterior probability density function. After that, the solutions are divided in complexes, a process of mixture that assures the survival of the sequences through global information sharing, independently achieved by each parallel sequence.

The evolution process of the parallel sequences, above described, is repeated until the criterion of convergence ([36]) is met.

The Sequence Evolution Metropolis (SEM) algorithm is responsible for the generation of new solutions carried out from a proposed probability distribution using C^k and S^k in L iterations, as follows:

1. Set iteration t = 1;
2. Compute the mean (μ^k) and the covariance ($\sum^k$) of the parameters of complexes C^k and let $\theta^{(t)}$ be the current drawn in S^k;
3. Draw $Z \sim U(0,1)$;
4. Draw candidate points $\theta^{(t+1)} \sim N(\theta^{(t)}, \sum^k)$ until $\theta^{(t+1)}$ is within Θ (feasible candidate);
5. Compute the posterior probability density function $p(\theta^{(t+1)}|y)$ using points in C^k and current draw $\theta^{(t)}$ of S^k. For the multiobjective optimization, use the fitness (f_{t+1}) computed by equation (3.16);
6. Calculate ratio $\alpha = [\frac{p(\theta^{(t)}|y)}{p(\theta^{(t+1)}|y)}]^{\beta/p(\theta^{(t+1)}|y)}$ where β is a scaling factor and $p(\theta^{(t)}|y)$ is the posterior probability density function associated with the current drawn of S^k. For the multiobjective case, the fitness concept (equation (3.16)) is used instead of the posterior probability density function;
7. If $Z > \alpha$, $\theta^{(t+1)} = \theta^{(t)}$;
8. Add $\theta^{(t+1)}$ to S^k and replace worst member of C^k with $\theta^{(t+1)}$;
9. If $t < L$, set iteration $t = t + 1$ and return to 2.

The parameters of the SCEM algorithm, or its multiobjective version MOSCEM, are the size of population (s) and number of sequences and complexes (q).

3.3.2 Multiobjective Approach Using Pareto Dominance Criterion

From a multiobjective context, it is necessary the introduction of a new concept to replace the simple comparison between uni-objective different solutions, i.e., the Pareto dominance concept. This multiobjective approach is described below and the consequent problems due to the use of the Pareto dominance concept are dealt afterwards.

Let a multiobjective minimization problem given by:

$$\min f(\mathbf{x}) = [f_1(\mathbf{x}), f_2(\mathbf{x}), \cdots, f_M(\mathbf{x})] \tag{3.15}$$

where $f_i(\mathbf{x})$ is i^{th} of M objective functions and $\mathbf{x}$ is a feasible solution.

Looking at equation (3.15), two distinct solutions (u and v) can be related in the following manner ([34]):

- If $f_i(u) \leq f_i(v) | \forall i = 1, \cdots, M$ and $f_i(u) < f_i(v)$ for some i, then v is strictly dominated by u, represented by $u \prec v$;
- Or, if $f_i(u) \leq f_i(v) | \forall i$, v is said to be weakly dominated by u, represented by $u \preceq v$.

If u is not dominated by v, and v is not dominated by u, u and v are said to be non-dominated solutions. It is clear, then, that multiobjective problems have more than one optimal solution, and this set of solutions is called Optimal Pareto front or True Pareto front composed of non-dominated solutions by any other solutions. The Pareto dominance concept basically assigns rank 1 to the non-dominated solutions and removing them from contention, then finding a new set of non-dominated solutions, now ranked 2, and so forth.

Figure 3.1(a) shows in the parametric space (θ) the location of the minimum of two objective functions represented by A and B, the line that connects these minimum and that are part of the Pareto front, and the γ point, which represents a possible solution of the optimal front. Figure 3.1(b) shows the same elements in the objective function space. It should be noticed that the curve that connects A to B is tangent to the function contour lines in the parameter space.

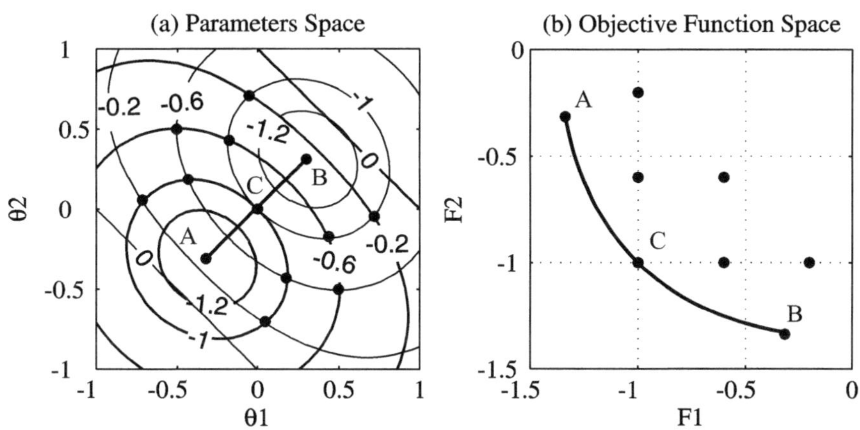

Fig. 3.1. Illustration of the Pareto optimal solution concept for a minimization problem with two objectives (F_1, F_2) ina search space ($\Delta\delta$) bi-parametric (θ_1, θ_2): a. parametric space; b. function space ([4])

Frequently, the number of solutions that belong to the Pareto front increases as the algorithm evolves, thus each non-dominated solution is a potential generator (queen) of new solutions in the next iteration of the algorithm. This would make the algorithm slower and more inefficient as the number of iteration increases, since each solution would generate a number of new solutions (new offspring and drones), escaping the user control.

In order to increase the algorithm efficiency, a clustering method ([37], [38]) was used to select the non-dominated solutions (queens) among the front solutions. The clustering technique used here not only promoted a better distribution of the solutions along the front but also improved the performance of the algorithm.

An alternative to the traditional Pareto dominance concept, Pareto-based ranking ([39]) correctly assigns all non-dominated individuals the same fitness, but this does not guarantee a uniformly sampled Pareto front. The paper by [4] made an improvement over the Pareto-based ranking proposed by [28]. This strategy is described below.

All non-dominated solutions are stored in a set *P'* and all other solutions stored in a set *P*. Each solution in *P'* is associated to a strength $r_i \in [0, 1]$, which represents the ratio between the number of dominated solutions of *P* dominated by i-th solution of *P'* and the total number of solutions of *P*. For all solutions in *P'*, the fitness is equals to the strength associated to it.

The fitness value for the solutions in *P* is given by the sum of the strengths of the solutions in *P'* that dominate them, summed by the traditional Pareto dominance rank and subtracted by 1 in order to obtain a fitness value greater than any other from solutions in *P'*:

$$f_j = \sum_{i=1}^{i \preceq j} r_i + rank_{Pareto} - 1 \tag{3.16}$$

where j is the i-th solution in *P*; r_i is the strength of the i-th solution in *P'* which dominates the j-th solution in *P*; and $rank_{Pareto}$ is the traditional Pareto dominance rank.

Through the use of the above concept it is possible to choose solutions that are on the Pareto front limits and to penalize those solutions that have many similar solutions in their neighborhood, preserving the diversity of the population. This favors solutions better distributed over the Pareto front, reducing the chances of clustering and premature convergence.

Multiobjective Honey-Bees Mating Optimization (MOHBMO)

In order to deal with multiobjective problems, some modifications in the HBMO algorithm were made. After the generation of the initial population (hive) and respective evaluation of objective functions, the selection of the "best solutions" (queens) must be made, but no longer based only on the comparison of single objective function values. Under a multiobjective approach, a new concept, such as the Pareto dominance concept, is needed for dealing with different solutions, classifying them as dominated or non-dominated solutions. The "best solutions" (queens) selected from the initial population are the non-dominated solutions.

Once identified the non-dominated solutions (queens), the iterative process is initiated in the same way as in the uni-objective case (mating flights, generation of new queens, improvement of the queens and of the new generation and selection of new queens). Each non-dominated solution (queen) will generate a certain number of solutions (offspring) after each iteration. The criteria for generation and improvement of the solutions (offspring) and of the best solution (queen) are the same employed in the uni-objective version.

With the new generated solutions (new offspring) and the non-dominated solutions from the previous iteration, the new set of non-dominated solutions is identified, which forms the Pareto front. These new solutions will generate the new solutions in the next iteration. The process is repeated until the stop criterion is satisfied.

Multiobjective Particle Swarm Optimization (MOPSO)

Many authors have recently proposed modifications in the PSO algorithm so it can be used in multiobjective frameworks ([40], [41], [42], [43], [34]). This study used the methodology proposed by [34].

The main difficulty of using PSO in multiobjective problems is how to select the components that guide the particles. In PSO, at each iteration, particles are modified according to both the best position that the particle experienced in the previous iterations and the best global position. If a new position of a particle is better than its best position up to this iteration, the best position of the particle is updated ([44]). In this case, the best position of a particle has no relation with other particles of the same population. However, within a multiobjective framework, in which the objective is to obtain a set of non-dominated solutions (particles) that form the Pareto front, it is mandatory that all particles share information with each other. Like in the MOHBMO described earlier, there is no clear definition of both best of the particle and global best solution.

The algorithm proposed by [34] basically consists of building a Pareto front at each iteration. The Pareto front is then updated at each iteration with the inclusion of the new set of dominant particles (solutions) and removal of dominated particles (solutions). This process is repeated until the maximum number of generations is reached. The algorithm starts with the random generation of a vector of positions of particles. At each iteration, one needs to evaluate if the new of position of each particle, obtained by equations (12) and (13), is dominated by the best position of that particle up to the current iteration. In case the new position is not dominated, the best position of the particle is updated.

In MOPSO, each particle has a best global solution associated to it. The selection of this best global solution is based upon a random selection from the solutions contained in the Pareto front in case the particle is part of the front. If the particle is not contained in the Pareto front, the random selection is made among all particles that are dominant.

Multiobjective Shuffled Complex Evolution Metropolis (MOSCEM)

A Markov Chain Monte Carlo sampler based on the SCEM algorithm, presented before, was proposed by [4] for solving multiobjective problems, in particular, for calibration of hydrologic models.

The MOSCEM algorithm employs the same evolving strategy of your uniobjective version SCEM, beside the use of the fitness ratio instead of the posterior probability ratio, and the objective functions instead of the posterior probability function, as described in section 3.1.

The initial steps of the MOSCEM algorithm are basically the same of the SCEM algorithm, say, generation of initial population, ranking solutions and parallel sequences initialization. The other steps are as follows:

1. The s points of D are, then, partitioned into q sets, said to be complexes, $C^q = D[qx(j-1)+q, 1:n]$ with j=*1,2,...,*m, where m is the number of solutions in each complex;
2. SCEM algorithm;
3. Unpack all Complexes C back into D and sort the points in increasing order of fitness value;
4. If stop criterion is met, stop, otherwise, return to step 1.

3.4 Evaluation of the Algorithms with Test Functions

3.4.1 Test Functions and Theirs Theoretical Minima

The HBMO, MOHBMO, PSO, MOPSO and MOSCEM algorithms were tested with mathematical functions which represent a challenge for any optimization algorithm, named here simply test functions ([29]). Five test functions were used to evaluate the optimization algorithms, noted here as $f_1(x_1,x_2), f_2(x_1,x_2), ..., f_5(x_1,x_2)$.The used functions are described in Table 3.1, but some of them are used only for the multiobjective case (for example, Function 1). Five combinations of functions in Table 3.1 were considered for the composition of the multiobjective problems (MO): (a) MO1: f_1 and f_2; (b) MO2: f_1 and f_3; (c) MO3: f_1 and f_4; (d) MO4: f_1 and f_5; and (e) MO5: f_3 and f_5. The difficulties imposed by those problems to any optimization algorithm are described in 3.1. To illustrate the difficulty involved, Figure 3.2 shows the behavior of functions 3 and 5, which correspond to MO5, as well as the optimal Pareto front and its location in the search space.

Sensitivity Analysis of the Algorithms' Parameters

A sensitivity analysis of the parameters of the aforementioned algorithms was performed in order to determine their values to be later used in the calibration of the hydrologic model. This sensitivity analysis is not presented here, but can be found in [45].

Table 3.1. Test functions in the uni- and multiobjective optimization

Function	Observation
$f_1(x_1,x_2)=x_1$	$for\ 0\leq x_1,x_2\leq 1$
$f_2(x_1,x_2)=-20\frac{sin(0.1+\sqrt{(x_1-4)^2+(x_2-4)^2})}{0.1+\sqrt{(x_1-4)^2+(x_2-4)^2}}$	$for\ -10\leq x_1,x_2\leq 20$
Difficulty: The function has several minimum and maximum, with global minimum at $x_1=4$ and $x_2=4$ equals to $f_2=-19.6683$.	
$f_3(x_1,x_2)=\frac{2-\exp\left[-\left(\frac{(x_2-0.2)}{0.004}\right)^2\right]-0.8.\exp\left[-\left(\frac{(x_2-0.6)}{0.4}\right)^2\right]}{x_1}$	$for\ 0.1\leq x_1\leq 1$ $for\ 0\leq x_2\leq 1$
The following expression represents a bimodal function, with a local minimum at $x_1=1$ and $x_2=0.6$ equals to $f=1.2$, and a global minimum at $x_1=1$ and $x_2=0.2$ equals to $f=0.7057$. Difficulty: Singularity at the global minimum.	
$f_4(x_1,g)=1-\left(\frac{x_1}{g}\right)^2$ $g(x_{m+1},\cdots,x_N)=g_{min}+$ $+(g_{max}-g_{min}).\left(\frac{\sum_{i=m+1}^{N}x_i-\sum_{i=m+1}^{N}x_i^{min}}{\sum_{i=m+1}^{N}x_i^{max}-\sum_{i=m+1}^{N}x_i^{min}}\right)^\gamma$	$for\ 0\leq x_1,x_2\leq 1$ $m=1;N=2$ $g_{min}=1;g_{max}=2$ $\gamma=0.25$
x_1^{min} and x_1^{max} are, respectively, the minimum and maximum values of the variable x_i, g_{min} and g_{max} are, respectively, the minimum and maximum values of the function $g(.)$. Difficulty: γ is the parameter responsible for the introduction of bias in the function values.	
$f_5(x_1,x_2)=(1+10x_2).$ $\left[1-\left(\frac{x_1}{1+10x_2}\right)^\alpha-\frac{x_1}{1+10x_2}\sin(2\pi q x_1)\right]$	$for\ 0\leq x_1,x_2\leq 1$ $for\ q=4;\alpha=2$

Uni-objective Minimization

In order to compare the performance between HBMO and PSO for minimization, both algorithms were used to minimize several test functions. The goals were to access both the ability of each algorithm to identify the optimum and their speed of convergence in terms of number of evaluation needed to reach the optimum. The initial condition was kept the same for both algorithms to guarantee a fair comparison.

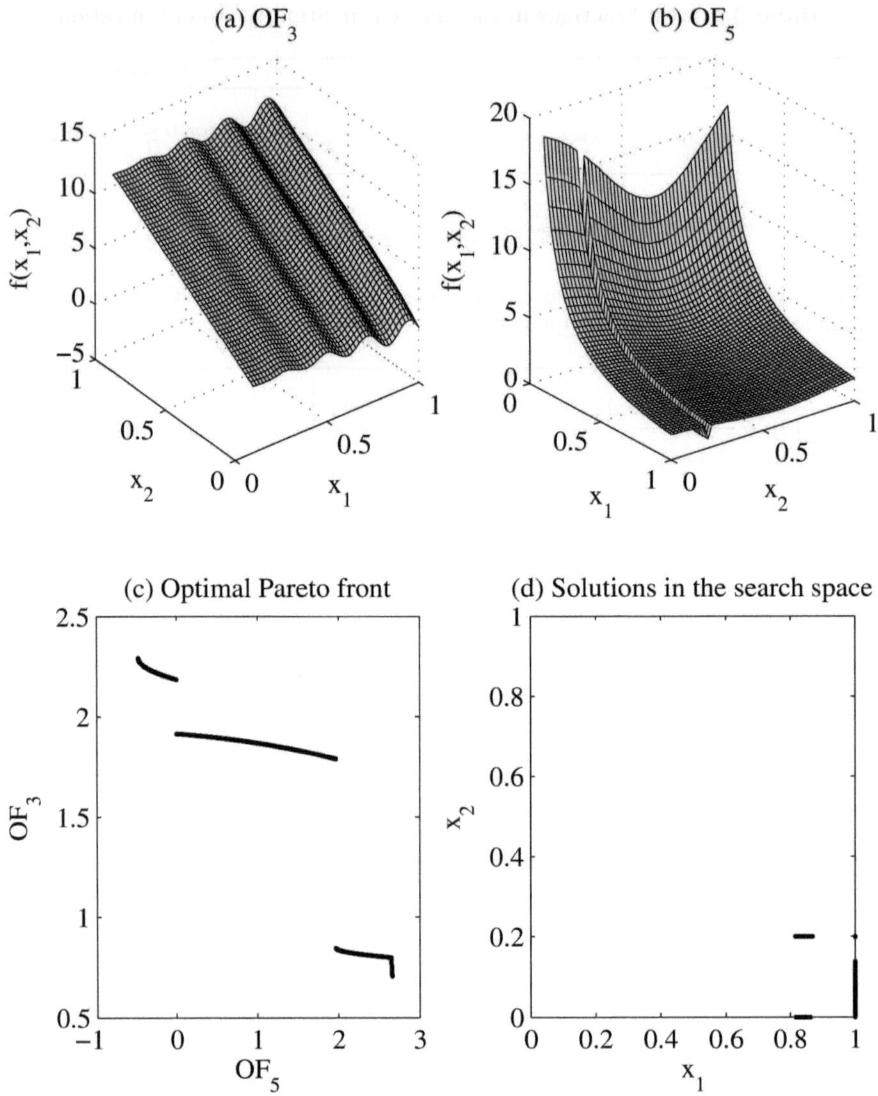

Fig. 3.2. Multiobjective Problem 5: Behavior of function 3 and 5 (a and b, respectivelly), Optimal Pareto front (c) and solutions in the search space (d).

For the uni-objective minimization of test functions, the following parameters of the HBMO algorithm were used: a. Initial population size = 100; b. Number of mating flights = 50; c. Number of queens = 20; d. Number of drones = 1; e. Minimum randomness factor of the drone = 10%; and, f. Number of offspring per queen = 4. For the PSO algorithm, the following parameters were used: a. Size of population = 100; b. Number of iterations = 50;

c. Maximum speed of the particle = 1 (function 4) and 0.1 (functions 2, 3 and 5); and, d. $c_1 = c_2 = 1$ and w varying from 0.95 to 0.4 until the algorithm reaches 70% of the maximum number of iterations, keeping the smallest value in the following iterations.

Once defined the algorithms' parameters, five applications were then made starting from the same initial conditions and parameter values for both algorithms as mentioned before. For each function, convergence was verified using graphics that show the value of the objective function at each iteration:

1. Function 2: Both algorithms have similar convergence with function values almost equal at the end of 700 iterations;
2. Function 3: PSO converged slower than HBMO. While HBMO rapidly reached the global optimum by 1500th evaluation, PSO did not identify the global minimum;
3. Function 4: For this problem, it was noted that PSO was more robust than HBMO, since PSO identified the global optimum for all cases but one, while HBMO identified the global optimum for two cases only;
4. Function 5: Although HBMO converged faster than PSO, both algorithms identified easily the global optimum.

MOHBMO and MOSCEM algorithms experienced problems in identifying the Pareto front in the presence of singularities (MO2), while MOPSO was able to fill in the front adequately. With respect to the presence of bias in the objective function (MO3), MOHBMO algorithm had the best performance and was the only algorithm to properly fill the Pareto front. For the multiobjective problem 4 (MO4), all three algorithms had similar performance, while for the multiobjective problem 5 (MO5) MOHBMO was able to better identify and adequately fill the Pareto front relative to the other two algorithms.

Multiobjective Optimization

The comparative evaluation of the algorithms in multiobjective problems employs multiobjective versions of the algorithms tested in the last section: the algorithms MOHBMO, MOPSO and MOSCEM, the multiobjective versions ofHBMOO, PSO and SCEM, respectively.

For the minimization of the multiobjective problems, the following parameters for MOHBMO algorithm were used: a. Initial population size = 100; b. Number of mating flights = 100; c. Number of queens = 20; d. Number of drones = 1; e. Minimum randomness factor of the drone = 1%; and, f. Number of offspring per queen = 4. The other parameters were used as described previously in section 5.1.1. For MOPSO algorithm, it was employed the following parameters a. Size of population = 100; b. Number of iterations = 100; c. Maximum speed of the particle = 0.5. The other MOPSO parameters were those recommended by [17]. For MOSCEM, it was used the following parameters: a. Size of population = 100; b. Number of complexes = 2; c. Number of Objective Function evaluations = 10,000.

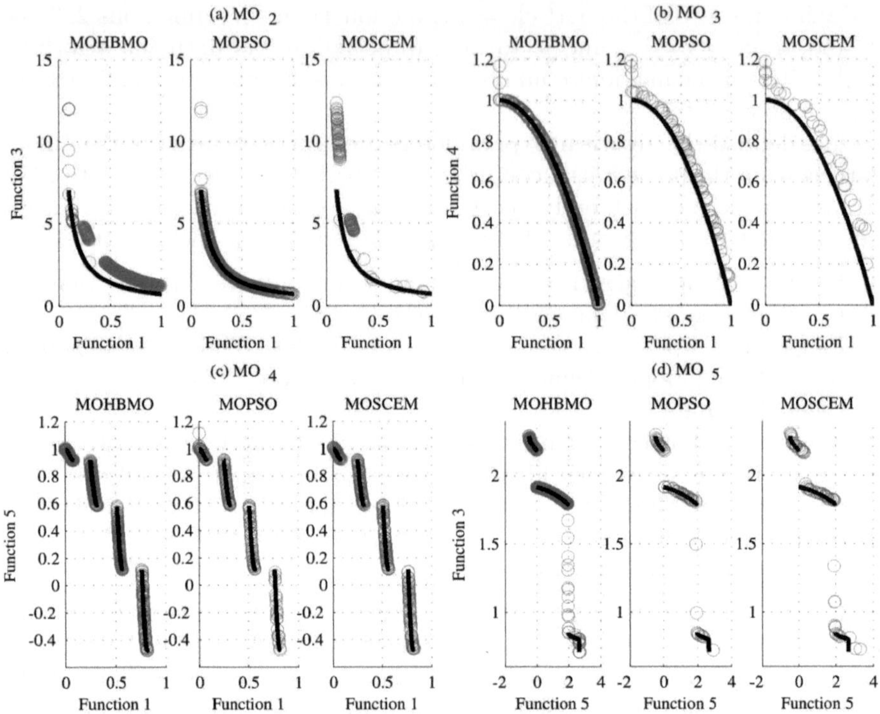

Fig. 3.3. True Pareto fronts and those identified by the algorithms MOHBMO, MOPSO and MOSCEM for the multiobjective problems 2–5.

Figure 3.3 presents the true Pareto fronts and those identified by the three algorithms MOHBMO, MOPSO and MOSCEM for the multiobjective problems 2-5. This figure shows that the algorithms PSO and MOPSO had better performance when compared to the other algorithms in the presence of singularities of the objective function. Also, HBMO and MOHBMO were superior to PSO and MOPSO/MOSCEM respectively, in terms of appropriate filling of the Pareto front and identification of the elements of the Pareto front in their limits. When bias was introduced in the objective function, PSO and MOPSO performed worse.

3.5 Use of Multiobjective Evolutionary Algorithms in Calibration of Hydrologic Models

This section evaluates the performance of the HBMO and MOHBMO algorithms relative to other relatively well known evolutionary algorithms (PSO, MOPSO, MOSCEM). The MOHBMO, a multiobjective version of the HBMO algorithm, was proposed in this chapter for the calibration of the hydrologic model HYMOD for stream gauge data recorded in the semi-arid region of the Brazilian Northeast.

The difficulty in calibrating daily hydrologic models for semi-arid regions is due mainly to the poor quality of the available hydrologic data as well as the inexistence of information regarding small reservoirs, which can affect tremendously in the streamflow generation process, in particular for low-flow periods. Another difficulty is the predominant convective nature of the precipitation regime over the rainy season, which makes even more necessary the existence of a dense monitoring network along with high quality information. This is very difficult to guarantee on a daily basis. The calibration study was carried out for gauge stream stations 34750000, 35760000 and 36125000 for the following periods 1989-1996, 1970-1973 and 1982-1989, respectively. For the validation study, when the calibrated set of parameters are tested against an independent observed series, the periods used were 1965-1988, 1967-1969 and 1968-1981, respectively. The use of a long record for calibration makes difficult the identification of reliable parameters ([46]), but the goal of the study was to evaluate the performance of the optimization algorithms. The Nash-Suttcliffe coefficient was employed as fitting criteria applied to daily streamflow series (of_1), characteristic points of the flow-duration curve (of_2), peak flows (of_3) and monthly volume series (of_4):

$$of = \max_{\theta}\left\{1 - \frac{\left(\sum_{i=1}^{N}(Q_i - \widehat{Q_i}(\theta))\right)^2}{\left(\sum_{i=1}^{N}(Q_i - \overline{Q})^2\right)}\right\} \tag{3.17}$$

where θ is the model parameter vector, Q_i is the observed streamflow variable (streamflows, flow duration curve values, peak flows), $\widehat{Q_i}$ is the simulated streamflow variable, $\overline{Q}$ is the streamflow variable mean and N is the size of the streamflow variable vector.

3.5.1 Hydrologic Model

The hydrologic model employed in this study was the daily rainfall-runoff model HYMOD ([47]). The used version is a relatively simple model, which uses the probability distribution concept to describe the spatial variation of runoff production process parameters. This allows the integration of the flow response over the whole watershed represented by algebraic expressions.

The idea underneath the model is that the watershed can be viewed as a set of points without interaction among them, while each one has a water storage capacity that, when exceeded, generates runoff. Figure 3.4 illustrates this representation.

The distribution function of the different water storage capacities is defined as:

$$F(C) = 1 - (1 - \frac{C}{C_max})^B \tag{3.18}$$

where F represents the cumulative probability of a certain water storage capacity (C) if a random point is selected; C_{max} is the largest water storage

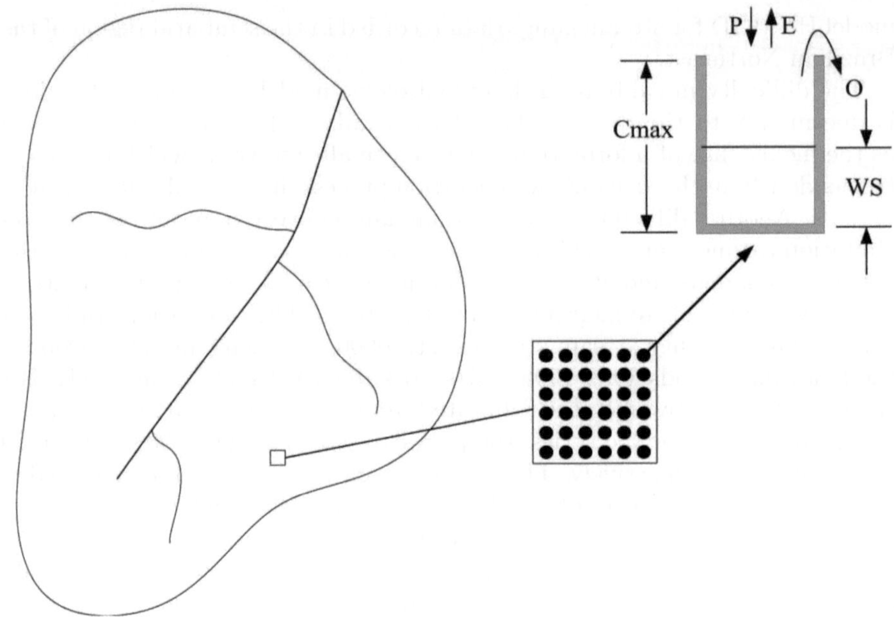

Fig. 3.4. Watershed representation in the HYMOD model (P: Precipitation; E: Evaporation; O: Outflow; WS: Water Storage; C_max is the largest water storage capacity within the watershed.

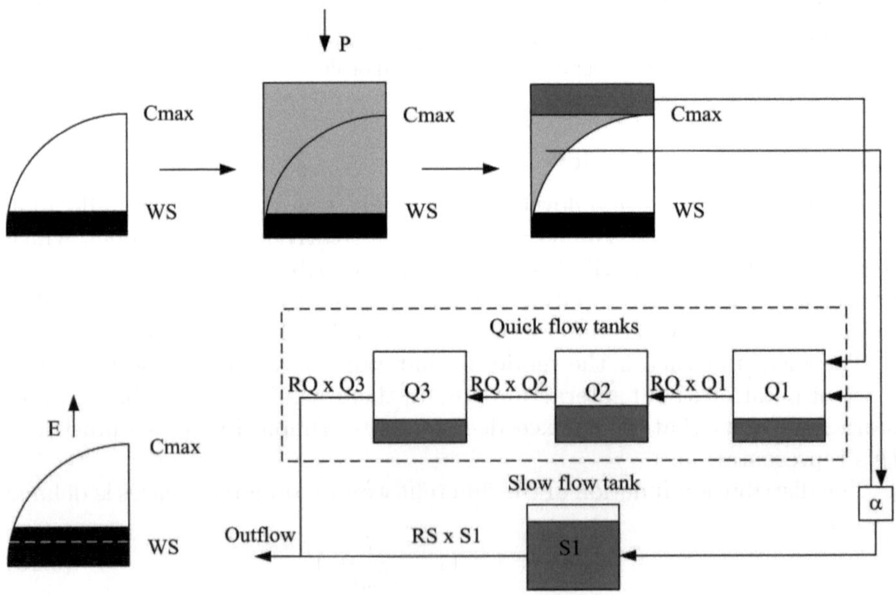

Fig. 3.5. Schematic representation of the hydrologic model HYMOD

capacity within the watershed and B is the degree of variability in the storage capacity. Figure 3.5 shows the schematic representation of the HYMOD model. After a rainfall event, the water can infiltrate up to the soil reaches its water storage capacity, after what runoff will be generated.

The fraction that exceeds C_{max} does not infiltrate and passes through three linear quick flow tanks at a constant flow rate RQ. For those points where the water storage is less than C_{max}, the remaining precipitation which exceeds the water storage is directed to either quick flow tanks or the slow flow tank depending on a constant α. The total outflow of the watershed is obtained by summing the outputs of the quick flow tanks and the slow flow tank.

Finally, the evaporation is taken from the water storage in the watershed. If the available water in storage is greater than the potential evaporation, the real evaporation is equal to the potential evaporation, otherwise all available water evaporates.

This model has five parameters: 1. Largest storage capacity within the watershed (C_{max}); 2. The degree of spatial variability in the water storage capacities (B); 3. Factor that divides the amount that exceeds the water storage capacity of points with a capacity lower than C_{max} between the

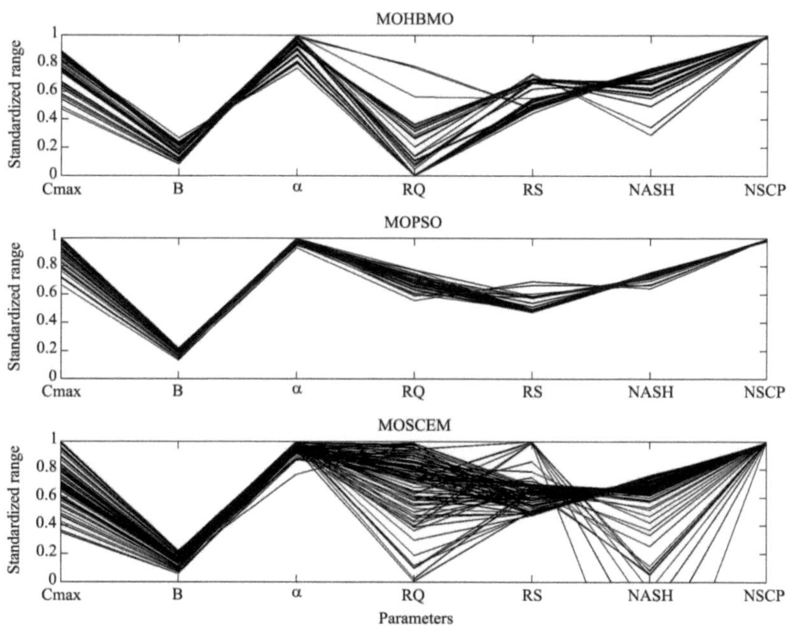

Fig. 3.6. Optimum solutions identified by the algorithms MOHBMO, MOSCEM and MOPSO using objective functions of_1 and of_2 in the calibration of the model HYMOD for the stream gauge station 34750000: Set of optimal parameters.

quick flow tanks and the slow flow tank (α); 4. Residence time for the quick flow tanks (RQ); 5. Residence time for the slow flow tank (RS).

3.5.2 Discussion of the Results

Figures 3.6 and 3.7 present the result for the calibration of HYMOD applied to the stream gauge station 34750000, selected from those stations that achieved better results among the 21 stations employed. The calibration was carried out for the 32 years of the station data, but only a period is shown here. This figure shows the sets of optimal parameters (Figure 3.6) and the Pareto fronts (Figure 3.7) identified by the multiobjective algorithms using objective functions 1 and 2. Generally, MOHBMO and MOSCEM were able to identify the Pareto front with a good density of points and coverage of its limits. The MOPSO experienced difficulty, as already noticed with the test functions, to identify points close to Pareto front limits with adequate density. For some stations, the Pareto front identified by the MOPSO was completely or partially dominated by the other Pareto fronts identified by MOHBMO and MOSCEM.

The observed hydrograph and those associated with the optimal solution set of the Pareto front identified by the MOHBMO algorithm are shown in Figure 3.8, again for the stream gauge station 34750000, but in this case,

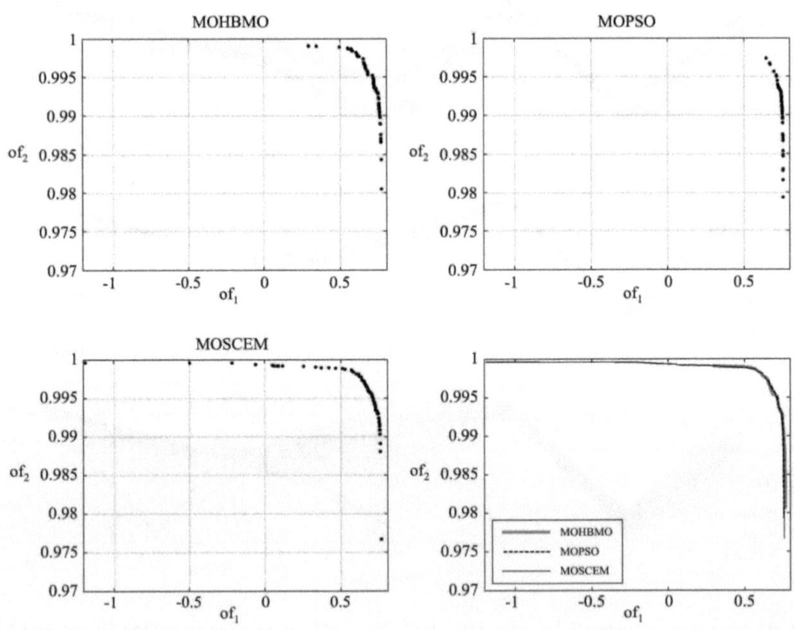

Fig. 3.7. Optimum solutions identified by the algorithms MOHBMO, MOSCEM and MOPSO using objective functions of_1 and of_2 in the calibration of the model HYMOD for the stream gauge station 34750000: Identified Pareto fronts.

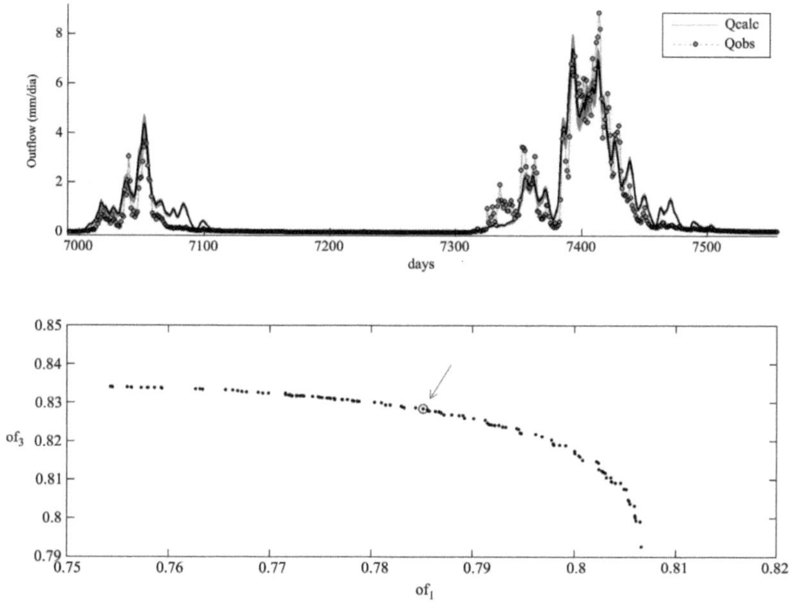

Fig. 3.8. Observed hydrograph and optimal hydrographs associated to Pareto front points applied to 34750000 stream gauge station using the objective functions of_1 and of_3. Dotted line represents the observed hydrograph while the continuous line represents the solution correspondent to the point of the Pareto front indicated by the arrow (Hydrographs associated to other points in the Pareto front are represented by grey lines). The identified Pareto front is presented below the hydrographs.

differently from Figures 3.6 and 3.7, the objectives are of_1 and of_3. The bold black line represents a trade-off solution between the two objectives. It should be noticed that the algorithms MOHBMO and MOSCEM did not experience any difficulty in determining the Pareto front for the case of highly correlated objectives and fractioned fronts. For these two algorithms the Pareto front filling was appropriate and extended to its range limits.

The algorithm MOHBMO was used to identify the Pareto front for the calibration period (PF_c) applied to 34750000 using the objective functions of_1 and of_2. The comparison of the observed hydrographs with those associated to the Pareto front solutions is show in Figure 3.9(a) and 3.9(b) for the calibration and validation period, respectively. In order to check how far those solutions would be if one use the validation period as the calibration period, the algorithm MOHBMO was again used to identify the Pareto front, now, for the validation period (PF_v).

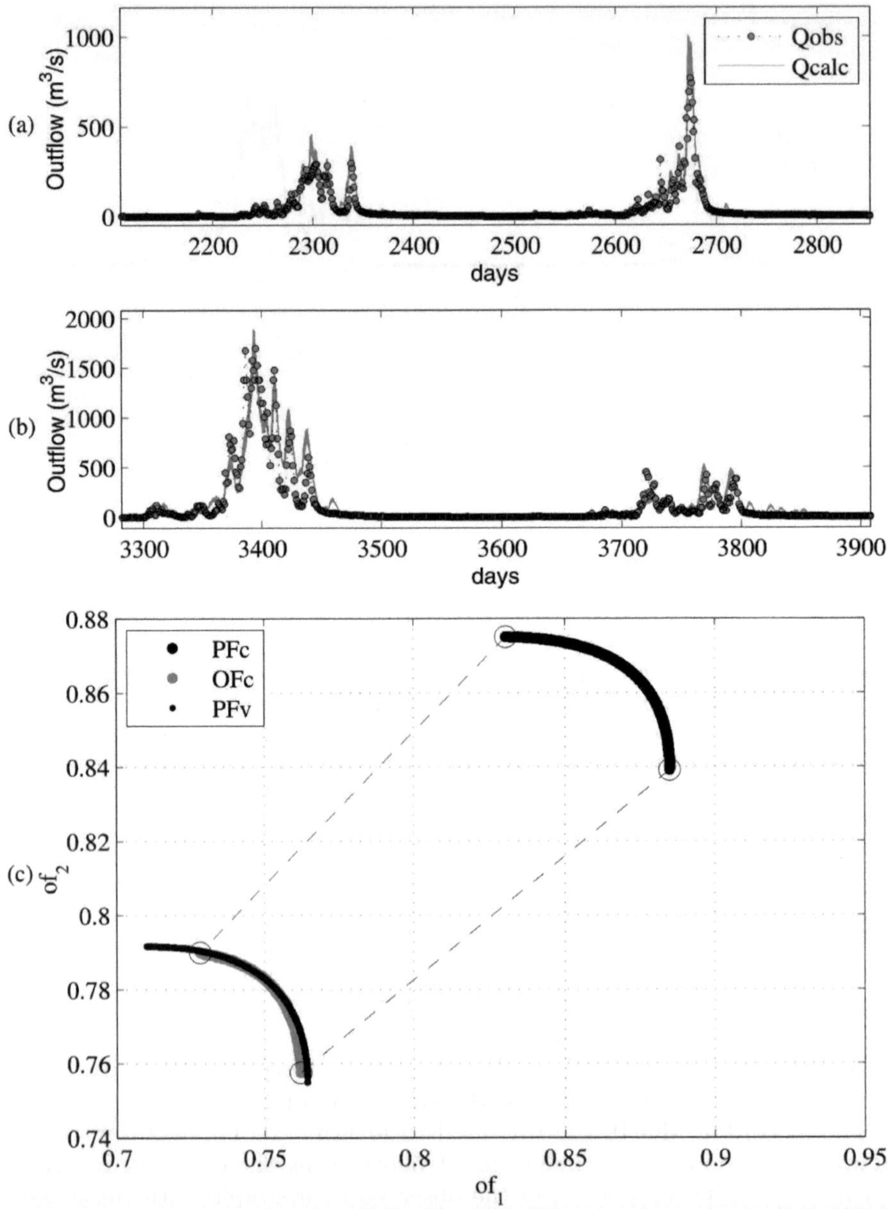

Fig. 3.9. Calibration and validation results for gauge station 34750000: (a) Observed and calibrated hydrographs; (b) Observed and validated hydrographs; (c) Identified Pareto fronts for the calibration and validation periods, and the objective function space for the validation perido using the calibrated parameters.

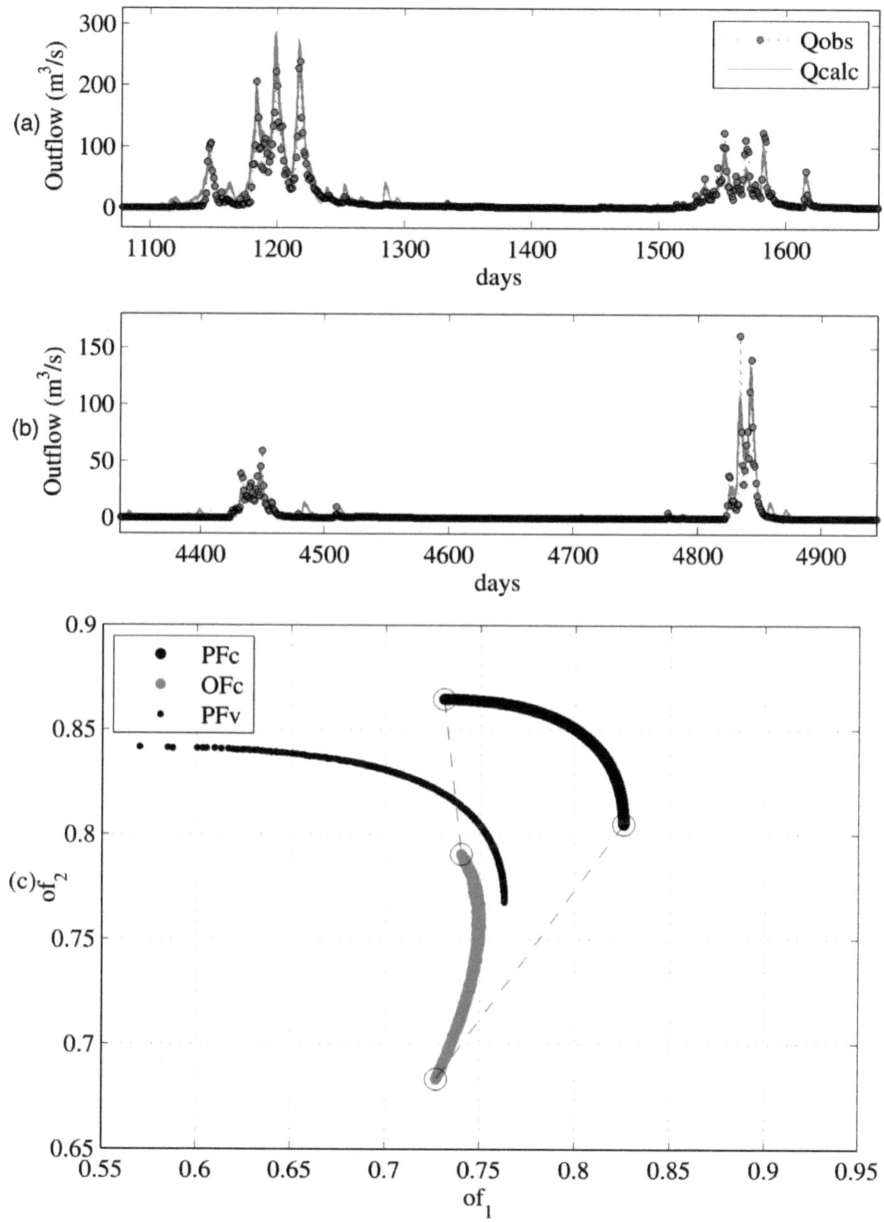

Fig. 3.10. Calibration and validation results for gauge station 36125000: (a) Observed and calibrated hydrographs; (b) Observed and validated hydrographs; (c) Identified Pareto fronts for the calibration and validation periods, and the objective function space for the validation perido using the calibrated parameters.

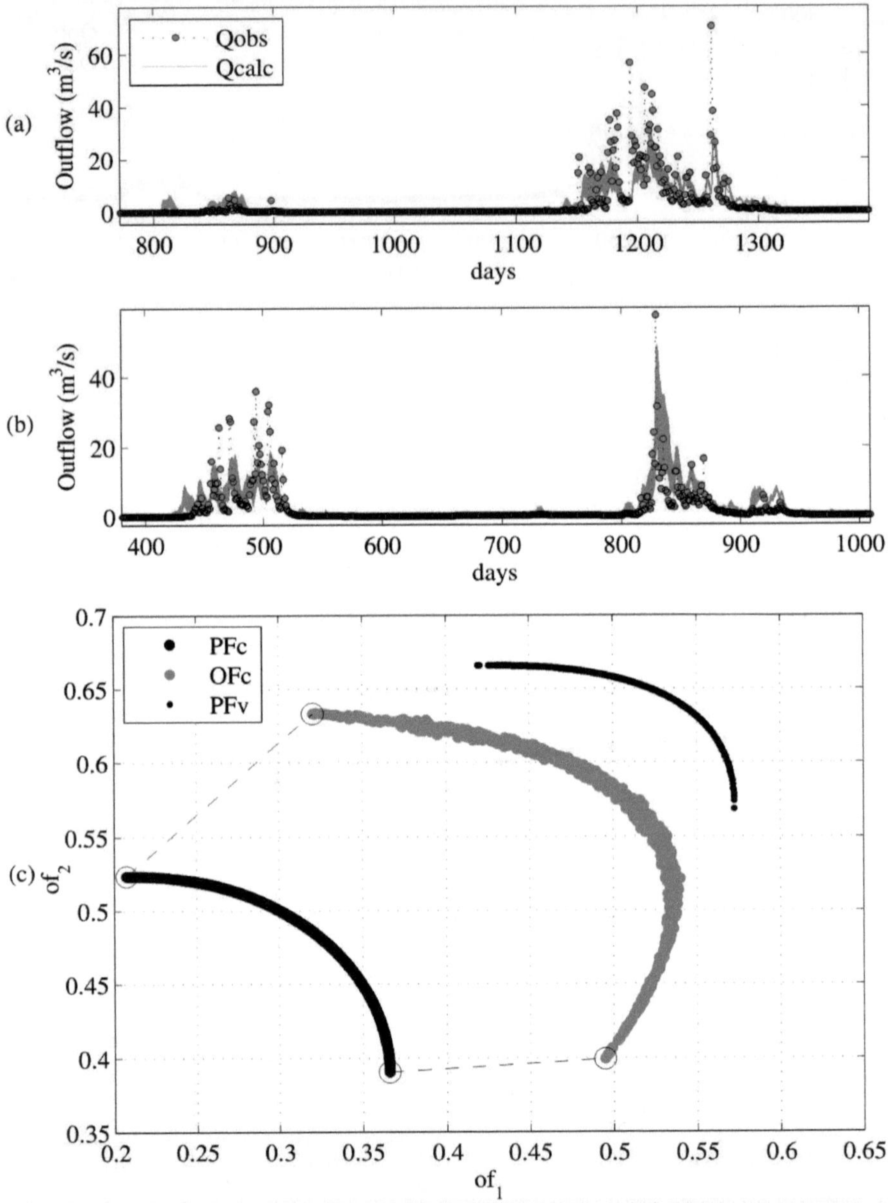

Fig. 3.11. Calibration and validation results for gauge station 35760000: (a) Observed and calibrated hydrographs; (b) Observed and validated hydrographs; (c) Identified Pareto fronts for the calibration and validation periods, and the objective function space for the validation perido using the calibrated parameters.

The parameters correspondent to PF_c were used to simulate the hydrologic behavior of the 34750000 watershed for the validation period. As a result, one can compute the objective function values (of_1 and of_2) with the PF_c solutions for the validation period and delineate a curve in the objective function space with these values (OF_c). Figure 3.9(c) shows the identified PF_v and the OF_c. The two curves are very close to each other and demonstrate that the PF_c solutions minimize both objectives of_1 and of_2 for the validation period as well, showing a stationary behavior of the hydrologic model parameters. For the PF_c and OF_c curves, the solutions that results in best values for either objective function 1 or 2 were marked with a circle and linked with a dotted line.

Figure 3.10 shows the same results for the gauge stream station 36125000. For this station, it is shown that the PF_v solutions are close to a small portion of the OF_c curve, which employs the PF_c solutions, but there is no PF_c solutions (gray curve) for of_1 small values. For station 35760000, it is shown that the PF_c solutions represent also a better choice for the validation period, since the OF_c curve "dominates" the PF_v completely (See Figure 3.11).

3.6 Use of Multiobjective Evolutionary Algorithms in Reservoirs' System Operation

This section illustrates the use of the multiobjective evolutionary algorithms MOHBMO and MOPSO to derive operating policies for the system of reservoirs that supplies water for the Metropolitan Region of Fortaleza (MRF), Ceará, located in the Brazilian northeast. The optimization study presented here employs two different objective functions, one related to the total pumping costs of the system, while the other is concerned with the total amount of water that is lost by evaporation. The trade-off between these two objectives obtained form this study is of great value for the water managers as they will be able to think more carefully about these issues.

3.6.1 Background

The Metropolitan Region of Fortaleza, with its 2.5 million inhabitants, is located in the state of Ceara, a semi-arid region of Brazil. The region obtains water for domestic and industrial purposes through the use of a complex system of reservoirs, which are linked by canals and water pump stations, bringing water from different basins within the state. The system consists of 5 reservoirs, 5 pumping stations, and a long canal (102 km) that diverts water from the Jaguaribe River basin, a large and agricultural basin in which the largest reservoirs of the State, with inter-annual storage capacity, are located.

Currently, the water management agency of the state (COGERH), responsible for the system, operates the system according to an operating policy that is based on a relatively simple set of rules bused upon current water storage in each reservoir. The goal is to improve the current policy taking into account two main objectives: (1) minimization of the pumping costs and (2) minimization of the amount of water losses through evaporation. The latter is justified by the extremely large potential evaporation rate observed in the region and the relatively different area-volume curves of these reservoirs.

3.6.2 The Reservoir System

The current system, with its five reservoirs, is linked to a much larger basin through Canal do Trabalhador, a canal that was built in 1993, during a severe drought period, to alleviate the water scarcity in the MRF and to reduce the risk of a collapse of the water supply system. The *Canal do Trabalhador* diverts water form the Jaguaribe River basin at the city of Itaiçaba, near the estuary of Jaguaribe, downstream all the reservoirs located in this basin. Figure 3.12 presents the system of reservoirs.

The operation of the current system assumes that the diverted flow from the Jaguaribe river basin will always be enough to meet MRF's demands, respected the capacity of both the pumping station at Itaiçaba and the *Canal do Trabalhador*. In this study there was no concern on how to operate the reservoirs in the Jaguaribe River basin in order to deliver the necessary amount of water to the MRF. There is a study in progress that is trying to perform a much larger optimization study, which considers not only the system of MRF, but also those reservoirs located in the Jaguaribe River basin. This study is important for planning purposes given the MRF's water demand is supposed to increase by almost 100% in the next 20 years.

The current system consists of five reservoirs: (1) Aracoiaba, with 170 hm^3 of storage capacity and drainage area of 584 km^2; located upstream of (2) Pacajus reservoir, whose drainage area is about 4,490 km^2, with storage capacity of 240 hm^3; (3) Pacoti reservoir, which drains an area of nearly 1,080 km^2, and that is linked through a canal to (4) Riachão, a small reservoir, whose drainage area is of just 34 km^2. Both Pacoti and Riachão reservoirs, which have jointly a storage capacity of 380 hm^3, are linked to (5) Gavião with 32.9 hm^3 of storage capacity and nearly 95 km^2 of drainage area.

The system has also five pumping stations that are used to divert water from one reservoir to another. The first one, named Itaiçaba Pumping Station, located near the city of Itaiçaba, is used to bring water from the Jaguaribe River basin to the Pacajus reservoir through Canal do Trabalhador. The Itaiçaba Pumping Station is able to divert up to 6 m^3/s to Canal do Trabalhador, which has at the moment a maximum flow rate of 5 m^3/s. Three other pumping stations, named PS0, PS1 and PS2, are used to bring water from the Pacajus reservoir to the Pacoti reservoir. PS1 operates only when the Pacajus water level is below 29.5 m. Between Pacoti and Riachão reservoirs there is also a pumping station that operates only when the water level at the Pacoti reservoir is below 36 m. Table 3.2 presents a summary of the pumping stations characteristics.

Fig. 3.12. Current reservoirs' system used for water supply of the metropolitan region of Fortaleza. The system consists of 5 reservoirs and 3 pumping stations. The *Canal do Trabalhador* diverts water from the Jaguaribe River Basin, near the city of Itaiçaba into the Pacajus reservoir (Source: COGERH).

Table 3.2. Pump stations' capacity

Pumping Station	Maximum pumping rate (m^3/s)
Itaiçaba	6.0
PS0	5.0
PS1	5.0
PS2	5.0
Pacoti	5.0

The system operates basically to supply the demands of the MRF, although small local demands, in the vicinity of the reservoirs, should also be met. The Gavião reservoir is responsible for supplying water to the MRF's water treatment plant. Therefore, the system is operated in such a way that Gavião should be able to deliver 8 m^3/s to the plant. Besides, both Pacajus and Aracoiaba need to meet local demands of 0.3 and 0.2 m^3/s, respectively.

3.6.3 The Current Operating Policy

The reservoirs' system is operated to meet domestic and industrial demands of the MRF. The current operating policy, employed by the water management agency of the state (COGERH), is based on a relatively simple set of rules that relates the amount of water that should be released/pumped from each reservoir based upon current water storage in each reservoir. When a reservoir is located downstream another reservoir, its release rule depends also on the current water storage at the reservoir located upstream.

In order to understand the operating policy, one needs to know how the system works. The Gavião reservoir is responsible for providing water to the MRF's water treatment plant given it is directly linked to the plant. The Gavião reservoir, in turn, receives water from Pacoti-Riachão reservoirs, which are considered in the optimization procedure as a single reservoir given they are connected through a canal. Pacoti-Riachão reservoirs, in turn, receive water from the Pacajus reservoir, which can be supplied either from Aracoiaba reservoir or *Canal do Trabalhador*, which brings water from the Jaguaribe River basin.

The current operating policy is summarized in Table 3.3. Columns Q+ and Q- indicate the amount of water that should flow into and be released or pumped from each reservoir, respectively. As can be seen, these releases depend on the current water storage. One can notice that the operating policy presented in Table 3.3 doesn't define rules for operating the Aracoiaba reservoir nor *Canal do Trabalhador*. This is the case because the operating policy considers both the Aracoiaba reservoir and *Canal do Trabalhador* as one system that releases water to Pacajus reservoir. Therefore, it does not specify for any given situation how much water should be brought by *Canal do Trabalhador*. It worth emphasizing that using *Canal do Trabalhor* to bring water to Pacajus implies in energy cost, while using Aracoiaba is cost free given that water flows by gravity. The rule applied in practice, and also employed in this study, is the following. In case the Aracoiaba reservoir is operating above 50% of its storage capacity, the water needed to be released to Pacajus reservoir is fulfilled by the Aracoiaba reservoir alone and nothing comes from *Canal do Trabalhador*. In case the Aracoiaba is operating below 25% of its storage capacity, the opposite occurs; the Pacajus reservoir receives water only from Canal do Trabalhador. Finally, in case the Aracoiaba water storage is within 25-50% of its storage capacity, 50% of the water released

Table 3.3. Current operating policy

				Pacajus Reservoir					
				> 50%		25-50%		< 25%	
		Q+	Q-	Q+	Q-	Q+	Q-	Q+	Q-
Pacoti/	> 50%	0.00	GAV	0.00	0.00	0.00	0.00	0.00	0.00
Riachão	25-50%	4.53	GAV	0.00	4.53	2.27	4.53	4.53	4.53
Reservoirs	< 25%	6.00	GAV	0.00	6.00	3.00	6.00	6.00	6.00

to Pacajus comes from Aracoiaba and the other 50% comes from *Canal do Trabalhador*.

Having said that, let us understand the operating policy provided in Table 3.3. For instance, if Pacoti-Riachão water storage is within 25-50% of its storage capacity and the Pacajus reservoir is also within 25-50% of its storage capacity, Pacoti-Riachão should release 8 m^3/s to Gavião in order to meet the MRF's demand. Besides, it should also receive 4.53 m^3/s from the Pacajus reservoir, which, in turn, should receive 2.27 m^3/s from either Aracoiaba reservoir or Canal do Trabalhador, or a combination of both. If Pacoti-Riachão water storage is above 50% of its storage capacity, it releases 8 m^3/s to Gavião, but it does not receive water from Pacajus reservoir, meaning the Pacajus reservoir doesn't need water from neither Aracoiaba reservoir nor *Canal do Trabalhador*. On the other hand, if Pacoti-Riachão is operating below 25% of its storage capacity, it still should release 8 m^3/s to Gavião reservoir, but it needs to receive 6 m^3/s from Pacajus. In this situation, the amount of water that the Pacajus reservoir should receive depends on its storage capacity, If it is operating above 50% of its storage capacity, it does not receive water from any source. However, if it is operating within 25-50% or below 25% of its storage capacity, it should receive 3 m^3/s or 6 m^3/s, respectively, which must come from either from Aracoiaba reservoir or Canal do Trabalhador.

3.6.4 Derivation of a New Operating Policy

The goal here is just to illustrate the use of multiobjective algorithms to derive optimized operating policies for reservoir operation. Therefore, the optimization study presented in the sequence is limited to obtaining a new policy based on the same structure of the current operating policy presented in Table 3.3. In other words, the optimization study carried out here obtains only new optimized values for releases and water storage ranges, but does not provide a new policy structure.

As it has been said before, the optimization study employed two objective functions: (1) minimize the total pumping costs of the system, and (2) minimize the total amount of water evaporated during the simulation period. A penalty function was included in both objective functions so as to guarantee that both MRF and local water demands are met.

The decision variables of this optimization study include nine release (or pumped) values, three values for each of the three reservoirs (Pacajus, Pacoti-Riachão, and Aracoiaba), and six water storage limit values, two for each reservoir, defining the three water storage ranges.

The five pumping stations of the system have different cost structures. Moreover, energy prices also vary seasonally. All these details were considered in the study. A more detailed description of the costs can be found in [45] and [48].

The study used a record of 85 years of inflow data available for all reservoirs. The last 60 years of data were used to derive the new operating policy of the system, while the first 25 years were employed to evaluate and compare the performance of both evolutionary algorithms and both operating policies, the current one used by the water management company of the State and the one obtained in this study.

Both algorithms were run with 1,000 and 10,000 evaluations of the objective functions. The MOHBMO algorithm used the following parameters: initial population size = 100; number of flights = 100 (1,000 for 10,000 evaluations); number of queens = 10; number of drones = 4; random factor of the drone = 10%; number of offspring per queen = 6. For the MOPSO, the following parameters were employed: population size = 100; number of iterations = 100 (for 1,000 evaluations) and 1,000 (for 10,000 evaluations). Ten different initial populations were used to obtain ten different Pareto fronts for each algorithm.

Results show that MOHBMO provided Pareto fronts with larger densities than those provided by MOPSO. In the case the maximum number of evaluations was set equal to 1,000, the average number of non-dominated solutions generated by MOHBMO was equal to 33, while the average number of solutions in the Pareto front obtained by MOPSO was 24. For 100,000 evaluations, the average number of non-dominated solutions was to 75 and 36 for MOHBMO and MOPSO, respectively.

Regarding the ability of both algorithms to find the optimum solution of each objective function, the results based on 100,000 evaluations of the objective function indicate that MOHBMO was able to identify better solutions in the extremes of the Pareto front than MOPSO. It means that MOHBMO generated solutions that provided the least pumping cost and the least amount of water lost by evaporation. When 10,000 evaluations were used, this conclusion is no longer valid.

Figure 3.13 presents a composite of 10 Pareto fronts obtained by both algorithms, MOHBMO and MOPSO, during the 60 year-optimization period, each front is associated with a different initial populations. One can see that MOHBMO, at least when 100,000 evaluations of the objective functions were allowed, provided better results than those obtained by MOPSO. MOHBMO was able to provide solutions with the least pumping cost and the least amount of water lost by evaporation. This result does not provide the final operating policy for the system given one still needs to pick

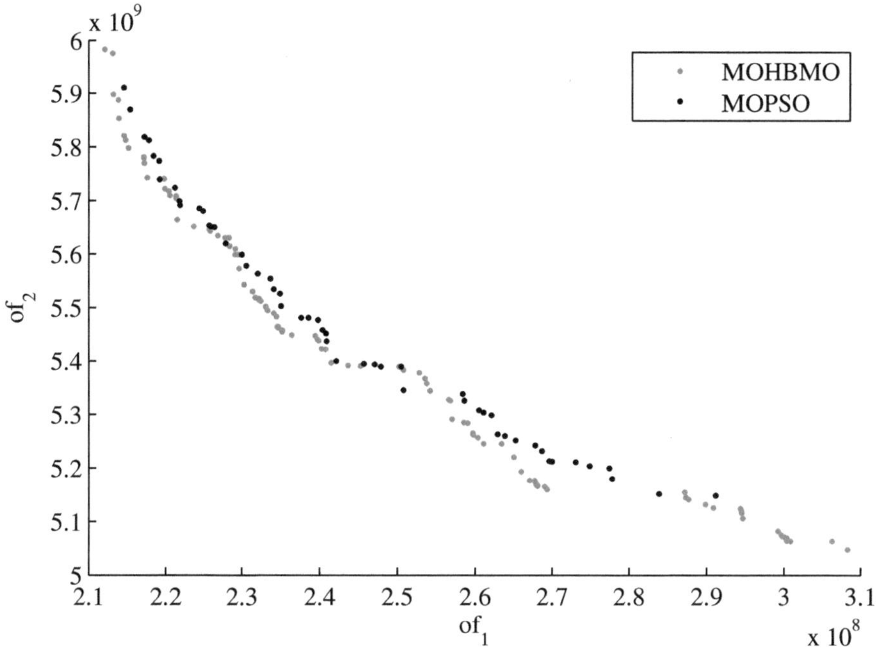

Fig. 3.13. Composite of 10 Pareto fronts obtained by MOHBMO and MOPSO with 10 initial populations and 100,000 evaluations (of_1 = pumping costs; of_2 = water evaporated).

a solution. Nonetheless, the Pareto front provides very interesting insights regarding both objectives. There is a strong feeling among water professionals that the water should be used in the most efficient way. Since the region presents a very large potential evaporation, water losses through evaporation is always a concern. This figure shows clearly that there is a cost for trying to avoid evaporation losses. This information is certainly valuable for those responsible for the operation of the system.

In order to compare the results obtained by this study with the results of the current operating policy, a minimum pumping cost solution was chosen. This solution indicates the operating policy presented in Table 3.4. It is important to point out that the optimization results indicate that there is no reason for the policy to have three different water storage zones for Pacoti-Riachão and Aracoiaba reservoirs. Both reservoirs should be operated based upon only two zones, as can be seen in Table 3.4.

Results based on simulation of the system for the 25 year-validation period shows that the optimized operating policy presented in Table 3.4 is better than the current policy presented in Table 3.3. The new operating policy obtained by MOHBMO provides a reduction of about 4% in total pumping costs during the validation period, and 5% reduction in the optimization

Table 3.4. Operating policy obtained by MOHBMO with minimum pumping cost. When Aracoiaba reservoir is above 24% of its storage capacity, only Aracoiaba supplies water to Pacajus, otherwise Canal do Trabalhador supplies water to Pacajus up to 5 m^3/s. If necessary, the remaining amount of water comes from Aracoiaba.

				Pacajus Reservoir					
				> 50%		25-50%		< 25%	
		Q+	Q-	Q+	Q-	Q+	Q-	Q+	Q-
Pacoti/	> 42%	0.01	GAV	0.45	0.01	0.74	0.01	3.34	0.01
Riachão	≤ 42%	5.60	GAV	1.30	5.60	5.06	5.60	5.18	5.60

period. This is an important reduction if one takes into account that the structure of the current operating policy has been preserved, and only the values of the releases and the ranges of water storages for each reservoir have been optimized. It is likely that a new structure of the operating policy can provide a more drastic reduction in costs and evaporation losses.

Besides, the differences between the current and the optimized policies can be evaluated by the number of times the system failed to meet the demand. The use of the current operating policy resulted in eight failures during the validation period and also eight failures during the optimization period. On the other hand, the use of the optimized policy resulted in only one failure during the validation period and none during the optimization one.

A different solution in the Pareto front, in the opposite direction of the minimization of pumping costs, is able to reduce the amount of water lost from evaporation by 16%, certainly not a negligible amount. However, the Pareto front indicates a clear trade-off between reducing pumping costs and reducing the amount of water evaporated from the reservoirs. No decision has been made to define the best strategy. The main result of the study is the Pareto front that provides the necessary elements for the managers of the water management company to think more deeply about these issues.

3.7 Summary

The optimization algorithms presented in this chapter were tested with mathematical functions for which the optima are known (optimum solution for the single-objective case and the Pareto fronts for the multiobjective case), and afterwards used in the calibration of the hydrologic model HYMOD and identification of optimal operating policies for the reservoir's system that supplies water for the Metropolitan Region of Fortaleza (MRF) and other smaller local demands.

For the test functions, the algorithms PSO and MOPSO had better performance when compared to the other algorithms in the presence of singularities of the objective function. In general, HBMO and MOHBMO were superior to PSO and MOPSO/MOSCEM respectively, in terms of appropriate filling of the Pareto front and identification of the elements of the Pareto front in their limits. The PSO and MOPSO performed worse when bias was introduced in the objective function.

The efficiency of the algorithm MOHBMO in identifying the Pareto front during the calibration of the HYMOD model was compared to those fronts identified by MOPSO and MOSCEM. In general, MOHBMO and MOSCEM presented a better performance relative to MOPSO, but MOSCEM, besides the higher computing time when compared to the other two algorithms, offers additional information regarding the parametric uncertainty of the model. The choice of the objective function within a multiobjective optimization framework should be carefully made, since this choice can dramatically affect the dynamics of the hydrologic simulation output. It is necessary to take more advantage from the multiobjective optimization by exploring different characteristics of the observed hydrographs to be preserved.

The MOPSO and MOHBMO algorithms were also employed to derive new operating policies for the reservoir's system of the Metropolitan Region of Fortaleza, Brazil. This optimization study preserved the same decision structure of the current operating policy employed by the water management company of State (COGERH). Current operations are based on a relatively simple set of rules that define the releases of each reservoir based upon the current water storage in one or more reservoirs, depending on its specific location within the system.

The goal of this last case of study was to redefine the values of the current operating policy in order to minimize both the total pumping costs and the amount of water lost by evaporation, while meeting the water demands from the Metropolitan Region of Fortaleza. Results show that the total pumping costs can be reduced by 4% when the minimum cost objective function is used to define the best operating policy. Since the structure of the original strict operational policy was preserved, leaving little room for improvement, this reduction is quite relevant. It is likely that a not so restrictive structure for the operating policy can provide a more drastic reduction in costs and evaporation losses. The MOHBMO algorithm provided Pareto fronts greater number of non-dominated solutions than those achieved by the MOPSO algorithm. Moreover, MOHBMO was able to identify solutions in the extremes of the Pareto front, while MOPSO did not.

The use of the multiobjective evolutionary algorithms for water resources engineering problems is very promising. This chapter tried to provide examples of use of such algorithms in this area to demonstrate their potential value.

Acknowledgements

This research was supported by Brazilian Funding Agencies CNPq and FINEP. The authors are also thankful to COGERH and FUNCEME for the data and information used in this chapter.

References

1. Loucks, D.P., Beek, E.V.: Water Resources Systems Planning and Management: An Introduction to Methods, Models and Applications. In: Studies and Reports in Hydrology. UNESCO Publishing, Turin (2005)
2. Oliveira, R., Loucks, D. P.: Operating Rules for Multireservoir Systems. Water Resources Research 33(4), 839–852 (1997)
3. Duan, Q.: Global Optimization for Watershed Model Calibration. In: Calibration of Watershed Models. AGU, Washington DC (2002)
4. Vrugt, J.A., Gupta, H.V., Bastidas, L.A., Bouten, W., Sorooshian, S.: Effective and Efficient Algorithm for Multiple Objective Optimization of Hydrological Model. Water Resources Research 39(8) (2003), doi:10.1029/2002WR001746
5. Duan, Q., Gupta, V.K., Sorooshian, S.: Effective and Efficient Global Optimization for Conceptual Rainfall-runoff Models. Water Resources Research 28(4), 1015–1031 (1992)
6. Duan, Q., Gupta, V.K., Sorooshian, S.: A Shuffled Complex Evolution Approach for Effective and Efficient Global Minimization. J. Optim. Theory Appl. 76(3), 501–521 (1993)
7. Duan, Q., Sorooshian, S., Gupta, V.K.: Optimal Use of the SCE-UA Global Optimization Method for Calibrating Watershed Models. Journal of Hydrology 158, 265–284 (1994)
8. Coello, C.A.C., Pulido, G.T., Lechuga, M.S.: Handling Multiple Objectives With Particle Swarm Optimization. IEEE Transactions on Evolutionary Computation 8(3), 256–279 (2004)
9. Reyes-Sierra, M., Coello, C.A.C.: Multi-objective particle swarm optimizers: A survey of the state-of-the-art. International Journal of Computational Intelligence Research 2(3), 287–308 (2006)
10. Das, I., Dennis, J.: A Closer Look at Drawbacks of Minimizing Weighted Sums of Objectives for Pareto Set Generation in Multicriteria Optimization Problems. Structural Optimization 14(1), 63–69 (1997)
11. Jin, Y., Okabe, T., Sendhoff, B.: Dynamic weighted aggregation for evolutionary multiobjective optimization: Why does it work and how? In: Spector, L., Goodman, E.D., Wu, A., Langdon, W.B., Voigt, H.-M., Gen, M., Sen, S., Dorigo, M., Pezeshk, S., Garzon, M.H., Burke, E. (eds.) Proceedings of the Genetic and Evolutionary Computation Conference (GECCO 2001). Morgan Kaufmann Publishers, San Francisco (2001)
12. Baumgartner, U., Magele, C., Renhart, W.: Pareto Optimality and Particle Swarm Optimization. IEEE Transactions on Magnetics 40(2), 1172–1175 (2004)
13. Hu, X., Eberhart, R.C., Shi, Y.: Particle swarm with extended memory for multiobjective optimization. In: Proceedings of the 2003 IEEE Swarm Intelligence Symposium. IEEE Service Center, Indianapolis (2003)
14. Schaffer, J.D.: Multiple Objective Optimization with Vector Evaluated Genetic Algorithms. In: Genetic Algorithms and their Applications: Proceedings of the First International Conference on Genetic Algorithms, Lawrence Erlbaum, Mahwah (1985)
15. Knowles, J.D., Corne, D.W.: The Pareto Archived Evolution Strategy: A New Baseline Algorithm for Multiobjective Optimization. In: 1999 Congress on Evolutionary Computation. IEEE Service Center, Washington D.C (1999)
16. Knowles, J.D., Corne, D.W.: Approximating the Nondominated Front Using the Pareto Archived Evolution Strategy. Evolutionary Computation 8(2), 149–172 (2000)

17. Nascimento, L.S.V., Reis Jr, D., Martins, E.: Avaliao do Algoritmo Evoluvionrio MOPSO na Calibrao Multiobjetivo do Modelo SMAP no Estado do Cear. In: Anais do XVII Simpsio Brasileiro de Recursos Hdricos, So Paulo (2007)
18. Xiao-Hua, Z., Hong-Yun, M., Li-Cheng, J.: Intelligent particle swarm optimization in multiobjective optimization. In: Congress on Evolutionary Computation (CEC 2005). IEEE Press, Edinburgh (2005)
19. Yapo, P.O., Gupta, H.V., Sorooshian, S.: Multiobjective Global Optimization for Hydrologic Models. Journal of Hydrology 204, 83–97 (1998)
20. Gupta, V.K., Sorooshian, S., Yapo, P.O.: Toward Improved Calibration of Hydrologic Models: Multiple and Noncommensurable Measures of Information. Water Resources Research 34(4), 751–763 (1998)
21. Bastidas, L.A., Gupta, V.K., Sorooshian, S., Shuttleworth, W.J., Yang, Z.L.: Sensitvity Analysis of a Land Surface Scheme using Multi-Criteria Methods. Journal of Geophysical Research 104(D16), 481–490 (1999)
22. Boyle, D.P., Gupta, H.V., Sorooshian, S.: Toward Improved Calibration Models: Combining the Strengths of Manual and Automatic Methods. Water Resources Research 36(12), 3663–3674 (2000)
23. Boyle, D.P., Gupta, H.V., Sorooshian, S.: Toward Improved Streamflow Forecasts: Value of Semi-distributed Modeling. Water Resources Research 37(11), 2749–2759 (2001)
24. Wagener, T., Boyle, D.P., Lees, M.J., Wheater, H.S., Gupta, H.V., Sorooshian, S.: A Framework for Development and Application of Hydrological Models. Hydrol. Earth Sys. Sci. 5(1), 13–26 (2001)
25. Fonseca, C.M., Fleming, P.J.: Genetic Algorithms for Multiobjective Optimization: Formulation, Discussion and Generalization. In: Forrest, S. (ed.) Proceedings of the Fifth International Conference on Genetic Algorithms. Morgan Kauffman Publishers, San Mateo (1993)
26. Horn, J., Nafpliotis, N., Goldberg, D.E.: A Niched Pareto Genetic Algorithm for Multiobjetive Optimization. In: Proceedings of the First IEEE Conference on Evolutionary Computation, IEEE World Congress on Computational Intelligence. IEEE Service Center, Piscataway (1994)
27. Srinivas, N., Deb, K.: Multiobjective Optimization Using Nondominated Sorting in Genetic Algorithms. Evolutionary Computation 2(3), 221–248 (1994)
28. Zitzler, E., Thiele, L.: Multiobjective Evolutionary Algorithms: A Comparative Case Study and the Strength Pareto Approach. IEEE Transactions on Evolutionary Computation 3(4), 257–271 (1999)
29. Deb, K.: Multi-objective Genetic Algorithms: Problem Difficulties and Construction of Test Problems. Evolutionary Computation 7(3), 205–230 (1999)
30. Haddad, O.B., Afshar, A., Marino, M.A.: Honey-Bees Mating Optimization (HBMO) Algorithm: A new heuristic approach for water resources optimization. Water Resources Management 20(5), 661–680 (2006)
31. Abbass, H.A.: A Pleometrosis MBO Approach to Satisfiability. In: Proceeding of International Conference on Computational Intelligence for Modeling, Control and Automation, CIMCA 2001, Las Vegas, USA (2001)
32. Lacerda, E.G.M.: Carvalho ACPLF, Introdução aos Algoritmos Genéticos. In: Galvão, C. O., Valença, M.J. S. (eds.) Sistemas Inteligentes: Aplicações a Recursos Hídricos e Ciências Ambientais. Ed. Universidade/UFRGS/ABRH, Porto Alegre (1999)
33. Kennedy, J., Eberhart, R.C.: Swarm Intelligence. Morgan Kaufmann Publishers, California (2000)

34. Alvarez-Benitz, J.E., Everson, R.M., Fieldsend, J.E.: A MOPSO algorithm based exclusively on pareto dominance concepts. In: Coello, C.A.C., Hernández Aguirre, A., Zitzler, E. (eds.) EMO 2005. LNCS, vol. 3410, pp. 459–473. Springer, Heidelberg (2005)
35. Thiemann, M., Trosset, M., Gupta, H., Sorooschian, S.: Bayesian Recursive Parameter Estimation for Hydrological Models. Water Resources Research 37(10), 2521–2535 (2001)
36. Gelman, A., Rubin, D.B.: Inference from Iterative Simulation Using Multiple Sequences. Stat. Sci. 7, 457–472 (1992)
37. Seber, G.A.F.: Multivariate Observations. Wiley, New York (1994)
38. Spath, H.: Cluster Dissection and Analysis: Theory, FORTRAN Programs, Examples, translated by J. Goldschmidt. Halsted Press, New York (1995)
39. Goldberg, D.E.: Genetic Algorithms in search, optimization and machine learning. Addison-Wesley, Reading (1989)
40. Coello, C.A.C., Lechuga, M.S.: MOPSO: A proposal for multiple objective particle swarm optimization. In: Congress on Evolutionary Computation (CEC 2002). IEEE Service Center, Piscataway (2002)
41. Hu, X., Eberhart, R.: Multiobjective optimization using dynamic neighborhood particle swarm optimization. In: Congress on Evolutionary Computation (CEC 2002). IEEE Service Center, Piscataway (2002)
42. Parsopoulos, K.E., Vrahatis, M.N.: Particle swarm optimization method in multiobjective problems. In: Proceedings of the 2002 ACM Symposium on Applied Computing (SAC 2002). ACM Press, Madrid (2002)
43. Fieldsend, J.E., Singh, S.: A multiobjective algorithm based upon particle swarm optimization, an efficient data structure and turbulence. In: Proceedings of the 2002 U.K. Workshop on Computational Intelligence, Birmingham, UK (2002)
44. Li, X.: A non-dominated sorting particle swarm optimizer for multiobjective optimization. In: Cantú-Paz, E., Foster, J.A., Deb, K., Davis, L., Roy, R., O'Reilly, U.-M., Beyer, H.-G., Kendall, G., Wilson, S.W., Harman, M., Wegener, J., Dasgupta, D., Potter, M.A., Schultz, A., Dowsland, K.A., Jonoska, N., Miller, J., Standish, R.K., et al. (eds.) GECCO 2003. LNCS, vol. 2724. Springer, Heidelberg (2003)
45. Barros, F.V.F.: The Use of Evolutionary Algorithms in the Calibration of Watershed Models and in the Optimization of Reservoirs' System Operation, MSc. Thesis, Department of Hydraulic and Environmental Engineering, Federal University of Ceara, Fortaleza, CE, Brazil (2007)(in Portuguese)
46. Yapo, P.O., Gupta, H.V., Sorooshian, S.: Automatic Calibration of Conceptual Rainfall-Runoff Models: Sensitivity to Calibration Data. Journal of Hydrology 181, 23–48 (1996)
47. Moore, R.J.: The probability-distributed principle and runoff production at point and basin scale. Hydrological Sciences Journal 30(2), 273–297 (1985)
48. FUNCEME, Optimization of the Reservoirs's System of the Metropolitan Region of Fortaleza to Minimize Pumping Costs, Report. Fortaleza, Brazil (2007)(in Portuguese)

4

Micro-MOPSO: A Multi-Objective Particle Swarm Optimizer That Uses a Very Small Population Size

Juan Carlos Fuentes Cabrera and Carlos A. Coello Coello*

CINVESTAV-IPN (Evolutionary Computation Group),
Computer Science Department, México
jfuentes@computacion.cs.cinvestav.mx, ccoello@cs.cinvestav.mx

In this chapter, we present a multi-objective evolutionary algorithm (MOEA) based on the heuristic called "particle swarm optimization" (PSO). This multi-objective particle swarm optimizer (MOPSO) is characterized for using a very small population size, which allows it to require a very low number of objective function evaluations (only 3000 per run) to produce reasonably good approximations of the Pareto front of problems of moderate dimensionality. The proposed approach first selects the leader and then selects the neighborhood for integrating the swarm. The leader selection scheme adopted is based on Pareto dominance and uses a neighbors density estimator. Additionally, the proposed approach performs a reinitialization process for preserving diversity and uses two external archives: one for storing the solutions that the algorithm finds during the search process and another for storing the final solutions obtained. Furthermore, a mutation operator is incorporated to improve the exploratory capabilities of the algorithm. The proposed approach is validated using standard test functions and performance measures reported in the specialized literature. Our results are compared with respect to those generated by the Nondominated Sorting Genetic Algorithm II (NSGA-II), which is a MOEA representative of the state-of-the-art in the area.

4.1 Introduction

The Particle Swarm Optimization (PSO) algorithm is a relatively recent heuristic based on the simulation of social behavior of birds within a flock [9]. Alhough PSO was originally proposed for balancing weights in neural networks, over the years, it has become a popular optimizer in a wide variety of

* The second author is also associated to the UMI-LAFMIA 3175 CNRS.

N. Nedjah et al. (Eds.): Multi-Objective Swarm Intelligent Systems, SCI 261, pp. 83–104.
springerlink.com

disciplines [14]. In the last few years, a variety of proposals for extending the PSO algorithm to handle multiple objectives have appeared in the specialized literature [27].

This chapter precisely proposes a new multi-objective particle swarm optimizer (MOPSO), called micro-MOPSO because of the very small population size that it adopts. Such a small population size combined with a good mechanism to preserve diversity allows us to produce reasonably good approximations of the Pareto front of several test problems of moderate dimensionality (up to 30 decision variables), while performing only 3,000 objective function evaluations. To the best of the authors' knowledge, this is the first micro-MOPSO ever proposed, and its main aim is to serve as a good alternative for applications in which only a very small implementation can be stored (e.g., when the multi-objective evolutionary algorithm needs to be placed into a microcontroller) or when only a low number of objective function evaluations can be afforded (e.g., in aeronautical engineering problems).

The organization of the rest of the chapter is the following. In Section 4.2, we define the problem of our interest. The Particle Swarm Optimization algorithm is introduced in Section 4.3. The previous related work is provided in Section 4.4. Section 4.5 describes our approach including the leader selection mechanism, the neighbors selection mechanism, the reinitialization process, and the mutation operator adopted. In Section 4.6, we present the performance measures, our experimental setup and the results obtained. Finally, Section 4.7 presents our conclusions and some possible paths for future research.

4.2 Basic Concepts

We are interested in solving problems of the type:

$$\text{Find } \mathbf{x} \text{ wich optimizes } \mathbf{f}(\mathbf{x}) = [f_1(\mathbf{x}), f_2(\mathbf{x}), \ldots, f_k(\mathbf{x})] \tag{4.1}$$

subject to:

$$g_i(\mathbf{x}) \leq 0 \quad i = 1, 2, \ldots, n \tag{4.2}$$

$$h_i(\mathbf{x}) = 0 \quad i = 1, 2, \ldots, p \tag{4.3}$$

where $\mathbf{x} = [x_1, x_2, \ldots, x_n]^T$ is the vector of decision variables, f_i, $i = 1, ..., k$ are the objective functions and g_i, h_j, $i = 1, ..., m$, $j = 1, ..., p$ are the constraint functions of the problem.

In multi-objective optimization problems the aim is to find good compromises (trade-offs). To understand the concept of optimality, we will introduce first a few definitions.

Definition 1. Given two vectors $\mathbf{x}, \mathbf{y} \in \mathbb{R}^k$, we say that $\mathbf{x} \leq \mathbf{y}$ if $x_i \leq y_i$ for $i = 1, ..., k$, and that **x dominates y** (denoted by $\mathbf{x} \prec \mathbf{y}$) if $\mathbf{x} \leq \mathbf{y}$ and $\mathbf{x} \neq \mathbf{y}$.

Definition 2. We say that a vector of decision variables $\mathbf{x} \in \mathcal{X} \subset \mathbb{R}^n$ is **nondominated** with respect to $\mathcal{X}$, if there does not exist another $\mathbf{x}' \in \mathcal{X}$ such that $\mathbf{f}(\mathbf{x}') \prec \mathbf{f}(\mathbf{x})$.

Definition 3. We say that a vector of decision variables $\mathbf{x}^* \in \mathcal{F} \subset \mathbb{R}^n$ ($\mathcal{F}$ is the feasible region) is **Pareto optimal** if it is nondominated with respect to $\mathcal{F}$.

Definition 4. The **Pareto Optimal Set** $\mathcal{P}^*$ is defined by:

$$\mathcal{P}^* = \{\mathbf{x} \in \mathcal{F} | \mathbf{x} \text{ is Pareto optimal}\}$$

Definition 5. The **Pareto Front** $\mathcal{PF}^*$ is defined by:

$$\mathcal{PF}^* = \{\mathbf{f}(\mathbf{x}) \in \mathbb{R}^k | \mathbf{x} \in \mathcal{P}^*\}$$

We thus wish to determine the Pareto optimal set from the set $\mathcal{F}$ of all the decision variable vectors that satisfy (4.2) and (4.3).

In general, when solving a multi-objective optimization problem with an evolutionary algorithm, we aim to produce as many elements of the Pareto optimal set as possible. Additionally, such solutions should be as uniformly distributed along the Pareto front as possible.

In spite of the existence of a variety of algorithms for solving multi-objective optimization problems in the operations research literature, such approaches normally become inefficient and even inappropriate when facing high-dimensional, discontinuous, and highly nonlinear problems [21]. Furthermore, there exist problems with a very large search space, which can also be very difficult to explore because of its shape. In all of these cases, the use of heuristic techniques is fully justified, since they will normally provide a reasonably good approximation of the Pareto optimal set within a reasonable time, although, without guaranteeing convergence [7]. One of such heuristic techniques is PSO, which is described next.

4.3 Particle Swarm Optimization

The Particle Swarm Optimization (PSO) algorithm was introduced by Russell Eberhart and James Kennedy in 1995 [9]. PSO is a population-based search algorithm based on the simulation of social behavior of birds within a flock [14, 10]. In PSO, each individual (particle) of the population (swarm) adjusts its trajectory according to its own flight experience and the flying experience of the other particles within its topological neighborhood in the search space.

In the PSO algorithm, the particles' positions and velocities are randomly initialized at the beginning of the search, and then they are iteratively updated, based on their previous positions and those of each particle's neighbors. Our proposed approach implements equations (4.4) and (4.5), proposed in [31] for computing the velocity and the position of a particle.

$$v_{id} = w \times v_{id} + c_1 r_1 (pb_{id} - x_{id}) + c_2 r_2 (lb_{id} - x_{id}) \tag{4.4}$$

$$x_{id} = x_{id} + v_{id} \tag{4.5}$$

where c_1 and c_2 are both positive constants, r_1 and r_2 are random numbers generated from a uniform distribution in the range [0,1], and w is the inertia weight that is generated in the range (0,1].

There are two versions of the PSO algorithm: the **global** version and the **local** version. In the global version, the neighborhood consists of all the particles of the swarm and the best particle of the population is called the "global best" (*gbest*). In contrast, in the local version, the neighborhood is a subset of the population and the best particle of the neighborhood is called "local best" (*lbest*).

4.4 Related Work

In order to extend the PSO algorithm to handle multi-objective optimization problems, the leader selection mechanism must be modified so that it incorporates the concept of Pareto optimality. Additionally, a mechanism for preserving diversity (normally called "density estimator") must be incorporated.

Moore and Chapman [22] introduced the first extension of PSO for handling multi-objective problems (MOPSO) in an unpublished manuscript. Since then, a wide variety of additional proposals have been introduced [12, 6, 23]. In [27], the authors present a survey of this area, in which a taxonomy of MOPSOs is also introduced. Based on that taxonomy, our approach can be classified as a Pareto-based MOPSO. Next, we will review the most representative Pareto-based MOPSOs that have been reported in the specialized literature, since they constitute the previous work related to our proposal.

In [12], Fieldsend and Singh proposed an approach that uses an elite external archive. This approach adopts a special data structure called "dominated tree" for storing the nondominated particles found along the search process. The archive interacts with the population in order to define leaders. This algorithm also adopts a "turbulence" operator that acts on the velocity value.

Coello Coello et al. [6] proposed a MOPSO that uses Pareto dominance to determine the flight direction of a particle. The algorithm uses a global archive for storing the nondominated solutions found during the search and also adopts a mutation operator that generates solutions within ranges of the decision variables that have not been previously used. In more recent work, Toscano Pulido and Coello Coello [34] proposed another MOPSO that divides the population into several swarms adopting clustering techniques.

Mostaghim and Teich [24] proposed another Pareto-based MOPSO, which adopts the so-called sigma method, in which the best local guides for each particle are adopted to improve the convergence and diversity. A mutation operator (called "turbulence") is also adopted in this case. In further work, Mostaghim and Teich [23] studied the influence of a relaxed form of Pareto dominance called ϵ-dominance [18] on MOPSO methods. In more recent work, Mostaghim and Teich [25] proposed a new method called *covering*MOPSO (cvMOPSO), which works in two phases. In the first phase, a MOPSO algorithm is run with a restricted archive size and the goal is to obtain a good approximation of the Pareto front. In the second phase, the nondominated solutions obtained from the first phase are considered as the input archive of the cvMOPSO. The particles in the population of the cvMOPSO are divided into subswarms around each nondominated solution after the first generation. The task of the subswarms is to cover the gaps between the nondominated solutions obtained from the first phase.

Li in [19] proposed an approach that incorporates the main mechanisms of the NSGA-II [8] into the PSO algorithm. In this algorithm, once a particle has updated its position, all the **pbest** positions of the swarm and all the new positions recently obtained are combined in just one set. Then, the approach selects the best solutions among them. The algorithm also randomly selects the leader based on two mechanisms: a niche count and a crowding distance.

In [3], Bartz-Beielstein et al. proposed an approach that starts from the idea of introducing elitism into a MOPSO, via an external archive. The authors analyzed different methods for selecting and deleting particles from the archive, aiming to generate a satisfactory approximation of the Pareto front.

Several other MOPSOs exist (see for example [38, 26, 1, 32, 35, 28]). However, none of them adopts a small population size, as our proposed approach. The use of small population sizes is unusual in the evolutionary algorithms literature in general, because its use normally speeds up the loss of diversity, and tends to generate premature convergence. However, in the genetic algorithms literature, it is known that the use of very small population sizes is possible, if an appropriate reinitialization process is adopted (such approaches are called micro-genetic algorithms (micro-GAs) [16, 4] and they use population sizes no larger than five individuals). Krishnakumar [16] proposed the first implementation of a micro-GA, and we are only aware of one multi-objective micro-GA, which was introduced in [5, 4]. This approach uses a population size of four individuals, and three forms of elitism: (1) an external archive that adopts the adaptive grid from the Pareto Archived Evolution Strategy (PAES) [15], (2) a population memory, in which randomly generated individuals are replaced by evolved individuals, and (3) a mechanism that retains the two best solutions generated by each run of the micro-GA.

In a further paper, Coello Coello and Pulido introduced the micro-GA^2 [33], which avoids the many parameters (eight) required by the original micro-GA for multi-objective optimization. The micro-GA^2 uses online adaptation and it performs a parallel strategy to adapt the crossover operator and the type of encoding (binary or real numbers) to be used.

To the authors' best knowledge, the approach described in this chapter is the first MOPSO in using a very small population size (i.e., a micro-MOPSO). The only micro-PSO that we are aware of is the (single-objective) micro-PSO for constrained optimization problems, which was developed by the same authors of this chapter [13].

4.5 The Micro-MOPSO

Our proposed micro-MOPSO is based on the global version of the PSO algorithm. It uses two external archives: one (called *auxiliary*) for storing the nondominated solutions that the algorithm finds throughout the search and another (called *final*) for storing the final nondominated solutions obtained. Our proposed algorithm performs the nondominated sorting introduced in [8] and uses a crowding distance for selecting leaders. As indicated before, our approach first selects the leader and then selects the neighborhood for creating the swarm. Our micro-MOPSO also uses a reinitialization process for preserving diversity and a mutation operator is incorporated to improve the overall exploratory capabilities of this heuristic (see Algorithm 1). The leader selection mechanism, the external archives together with the reinitialization process and the mutation operator adopted, are all described next in more detail.

Final archive

Since the micro-MOPSO uses a very small population size (only five particles), it needs an external archive for reporting the final solutions that it has found (this is called the *final archive*). Our algorithm uses this archive for selecting leaders (see Section 4.5.1). The upper bound of the final archive (FAB) is a parameter that needs to be set by the user.

At each iteration, and after updating the particles' positions, the nondominated solutions are obtained from the population of the micro-MOPSO. These solutions enter into the final archive and remain there only if no other solution dominates them during the entire evolutionary process. If a solution dominates another solution stored in the final archive, the stored solution is deleted. When the maximum capacity of the archive is reached, then the algorithm applies the crowding distance [8] as the criterion to prune the contents of the archive.

Algoritmo 1. Pseudocode of the Micro-MOPSO

Require: *Number of particles, number of generations, number of reinitialization particles (pr), number of reinitialization generations (gr), auxiliary archive bound (AAB), final archive bound (FAB), mutation rate.*

Ensure: *Nondominated solutions (final archive)*

```
Initialize the final archive (empty);
Initialize the auxiliary archive (empty);
for i = 1 to Number of particles do
   Initialize position and velocity randomly;
end for
Store the swarm into the auxiliary archive;
Get the set nondominated solutions;
Store the nondominated solutions into the final archive;
cont = 1;
repeat
   Select the leader;
   Select the neighborhood;
   if cont == number of generations for reinitialization then
      Reinitialization process;
      cont = 1;
   end if
   for i = 1 to Number of particles do
      Update the velocity;
      Compute the actual position;
      if x_i dominates to xpb_i then
         for d = 1 to Number of dimensions do
            xpb_id = x_id //Update the pbest;
         end for
      end if
   end for
   Performs mutation;
   Store the swarm into the auxiliary archive;
   if auxiliary archive length > AAB then
      Filter out the auxiliary archive;
   end if
   Get the nondominated solutions;
   Store the nondominated solutions into the final archive;
   if final archive length > FAB then
      Filter out the final archive;
   end if
   cont =cont + 1;
until Maximum number of generations
return the nondominated solutions (final archive)
```

Auxiliary Archive

The micro-MOPSO uses another external archive called *auxiliary archive*, which is used for storing the solutions that the algorithm finds along the search process. Our algorithm uses it for selecting the neighbor in the swarm. The upper bound of the auxiliary archive (AAB) is a parameter that needs to be set by the user.

At each iteration and after updating the particles' positions, the swarm is stored into the auxiliary archive. When the maximum limit imposed on the size of the archive is reached, the algorithm performs nondominated sorting for keeping the solutions located into the first five fronts or the best AAB solutions (see Figure 4.1).

4.5.1 Leader Selection

As we said before, the micro-MOPSO uses the final archive for selecting leaders. The idea is to have better diversity for the selection process. Algorithm 2 shows the mechanism for selecting a leader. The leader is selected from a subset of the final archive members that have the best crowding distances (i.e., the best spread). The size of the subset of leaders is a percentage (defined by the user) of the total population size.

Algoritmo 2. Pseudocode of the leader selection in micro-MOPSO

Require: *Final archive FA, final archive maximum limit (FAB), population percentage for selecting leader (PPS).*
Ensure: *Leader*
for $i = 1$ to Final archive length **do**
    Compute the crowding distance CD_i;
end for
Sort the final archive members (according to CD);
Get a percentage of particles (pps) from the population based on their CD values;
Choose a particle in a random way;
Make this particle the leader;

The use of the crowding distance for selecting leaders allows the micro-MOPSO to select nondominated solutions that are located in less crowded areas of the Pareto front. The use of diversity information allows us to discriminate from among several potential leaders (i.e., all the nondominated solutions produced so far). The potential disadvantage of this selection method is that, in the presence of local Pareto fronts, a good diversity maintenance mechanism is required, in order to avoid stagnation. In our case, our mutation

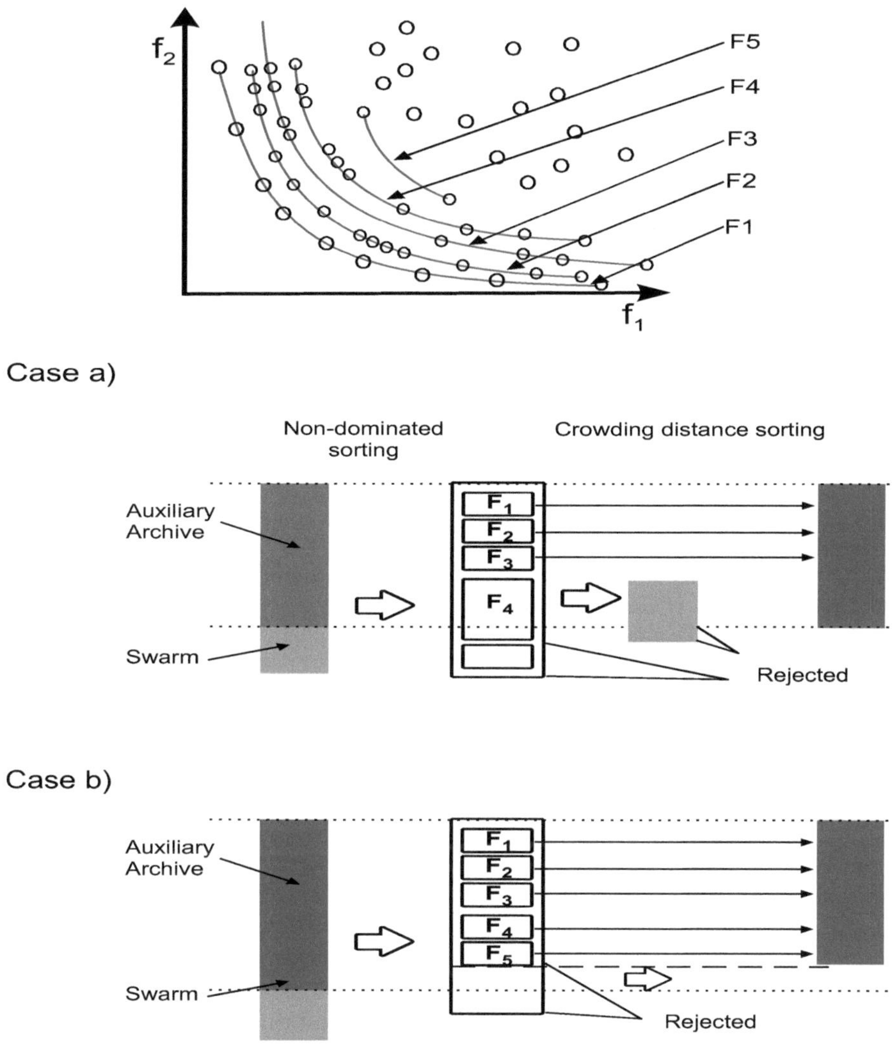

Fig. 4.1. An example in which the auxiliary archive exceeds its maximum allowable limit. The process finds the particles of the first nondominated front for all the archive members. If the front length is lower than the maximum limit, the front is kept into the archive. Then, in order to find the individuals in the next front, the solutions of the first front are temporarily disregarded and the above procedure is repeated until five fronts are found. Case a) When the front is kept into the archive and it exceeds the allowable limit, a crowding distance is computed (just for the front) in orden to filter out solutions, and the following fronts are deleted. Case b) The micro-MOPSO keeps into the auxiliary archive at most five fronts.

operator is responsible for such diversity maintenance, but other mechanisms could also be adopted (e.g., niching [29]).

Neighbors selection mechanism

In contrast with others MOPSOs, the micro-MOPSO first selects the leader and then uses the auxiliary archive for selecting the neighborhood for creating the swarm. Algorithm 3 shows the mechanism for selecting the neighbors which have the smallest Euclidian distance.

Algoritmo 3. Pseudocode for selecting the neighborhood

Require: *Leader gbest, auxiliary archive.*
Ensure: *Neighborhood*
1: **for** $i = 1$ to auxiliary archive length **do**
2: Compute the Euclidean distance (ED) from the ith particle to
3: the leader particle (*gbest*) in objective function space;
4: **end for**
5: Sort the auxiliary archive (according to the ED values);
6: Choose the $N-1$ particles closest to *gbest*;
7: Create a swarm with the $N-1$ particles chosen plus the leader (*gbest*);

This mechanism allows the micro-MOPSO to be explorative at the beginning of the search and exploitative at the end.

4.5.2 Reinitialization Process

The micro-MOPSO uses a reinitialization process similar to the one proposed in [13]. The mechanism is the following: after certain number of iterations (replacement generations **rg**), the algorithm identifies a certain number of nondominated solutions (**swarm**) and replaces them by randomly generated particles (**rp**). The rationale for mixing evolved and randomly generated particles is to avoid premature convergence.

4.5.3 Mutation Operator

Although the original PSO algorithm has no mutation operator, the addition of such a mutation operator is a relatively common practice nowadays in the specialized literature. The main motivation for adding this operator is to improve the performance of PSO as an optimizer, and to improve the overall exploratory capabilities of this heuristic [2]. In our proposed approach, we implemented the mutation operator originally developed by Michalewicz for genetic algorithms [20]. It is worth noticing that this mutation operator has been used before in PSO, but in the context of unconstrained multimodal optimization [11]. This operator varies the value added or substracted to a solution during the actual mutation, depending on the current iteration number (at the beginning of the search, large changes are allowed, and they become very small towards the end of the search). We apply the mutation operator in the particle's position, for all its dimensions:

$$x_{id} = \begin{cases} x_{id} + \Delta(t, UB - x_{id}) & \text{if } R = 0 \\ x_{id} - \Delta(t, x_{id} - LB) & \text{if } R = 1 \end{cases} \tag{4.6}$$

where t is the current iteration number, UB is the upper bound on the value of the particle's dimension, LB is the lower bound on the particle's dimension, R is a randomly generated bit (zero and one both have a 50% probability of being generated) and $\delta(t, y)$ returns a value in the range $[0, y]$. $\delta(t, y)$ is defined by:

$$\Delta(t, y) = y * (1 - r^{1-(\frac{t}{T})^b}) \tag{4.7}$$

where r is a random number generated from a uniform distribution in the range [0,1], T is the maximum number of iterations and b is a tunable parameter that defines the non-uniformity level of the operator. In this approach, the b parameter is set to 5 as suggested in [20].

4.6 Experiments and Results

When an evolutionary algorithm is used for solving multi-objective problems, three issues have to take into consideration for evaluating the performance of the algorithm [39]:

1. Minimize the distance of the Pareto front produced by the algorithm with respect to the true Pareto front.
2. Maximize the spread of solutions found, so that we can have a distribution of vectors as smooth and uniform as possible.
3. Maximize the number of elements of the Pareto optimal set found.

We adopt three performance measures for evaluating the performance of the micro-MOPSO and for comparing it with respect to the NSGA-II that is an algorithm representative of the state-of-the-art in the area.

4.6.1 Performance Measures

In order to assess the performance of our proposed approach and compare it with respect to the NSGA-II, we adopted three performance measures that have been commonly used in the specialized literature: error ratio, inverted generational distance, and spacing. Each of them is briefly described next.

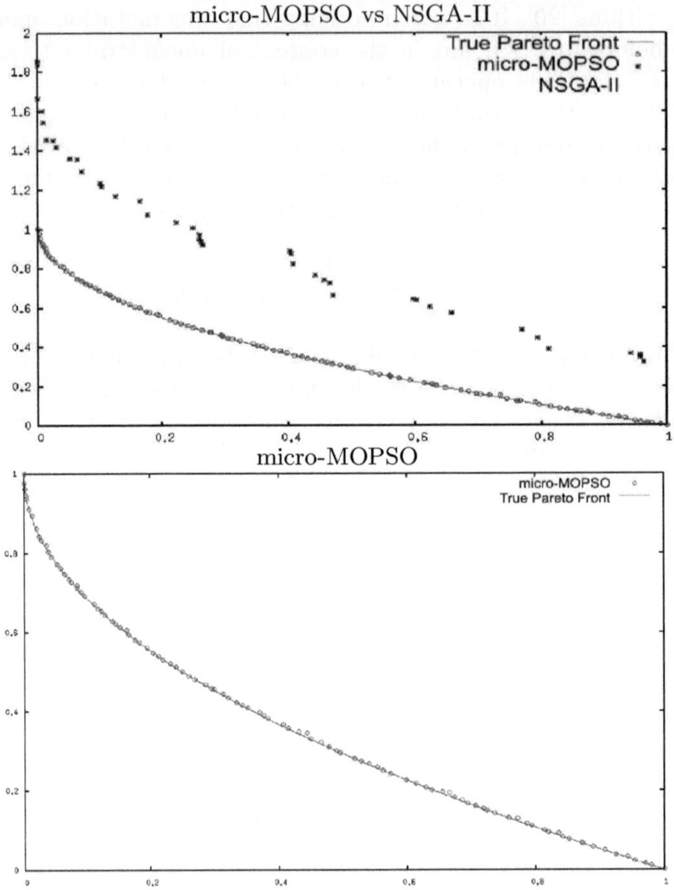

Fig. 4.2. Graphical comparison of the Pareto fronts found by our micro-MOPSO (bottom) and the NSGA-II (top) for ZDT1.

Error Ratio

The error ratio (ER) performance measure reports the number of vectors in the Pareto front found that are not members of the true Pareto front [36]. This is mathematically represented by:

$$ER = \frac{\sum_{i=1}^{n} e_i}{n} \tag{4.8}$$

$$e_i = \begin{cases} 1 \text{ if the } i\text{th vector is member of the true Pareto front} \\ 0 \text{ otherwise} \end{cases} \tag{4.9}$$

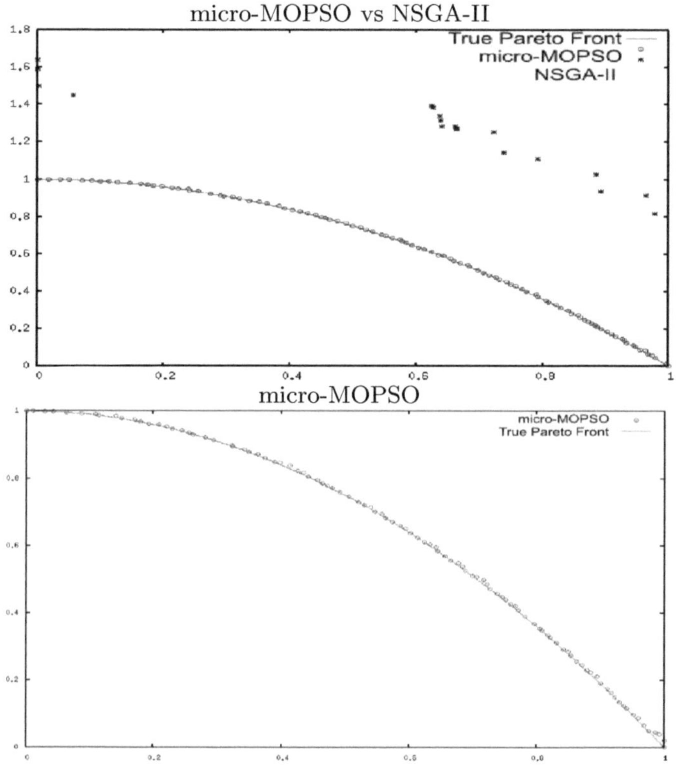

Fig. 4.3. Graphical comparison of the Pareto fronts found by our micro-MOPSO (bottom) and the NSGA-II (top) for ZDT2.

Inverted Generational Distance

The generational distance (GD) reports how far, on average, the Pareto front found by an algorithm is from the true Pareto front. Mathematically, it is defined by:

$$GD = \frac{(\sum_{i=1}^{n} d_p^i)^{1/p}}{n} \tag{4.10}$$

where n is the number of vectors that the algorithm found, for $p = 2$, d_i is the Euclidean distance (in objective space) between each member, i, of the Pareto front found and the closest member in the true Pareto front.

The inverted generational distance (IGD) reports how far, on average, the true Pareto front is from the Pareto front found by an algorithm. This intends to reduce some of the problems that occur with the original generational distance metric when an algorithm generates very few nondominated solutions.

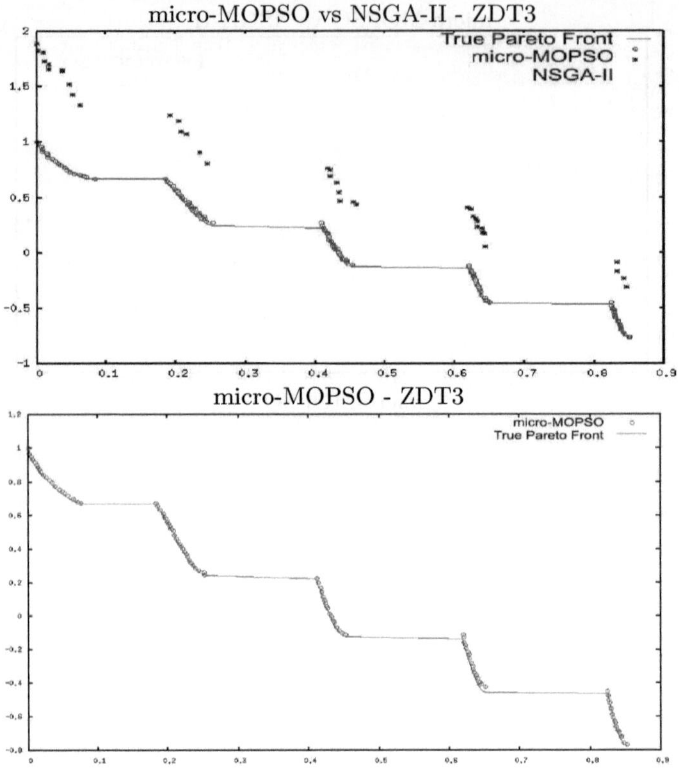

Fig. 4.4. Graphical comparison of the Pareto fronts found by our micro-MOPSO (right) and the NSGA-II (left) for ZDT3.

Spacing

The spacing (S) performance measure numerically describes the spread of the vectors in the Pareto front found by the algorithm [30]. It measures the distance variance of neighboring vectors in the Pareto front found. The equations 4.11 and 4.12 define this performance measure.

$$E = \sqrt{\frac{1}{n-1}\sum_{i=1}^{n}(\overline{d} - d_i)^2} \tag{4.11}$$

$$d_i = min_{i,j \neq i}(\sum_{m=1}^{M} |f_i^m - f_j^m|) \tag{4.12}$$

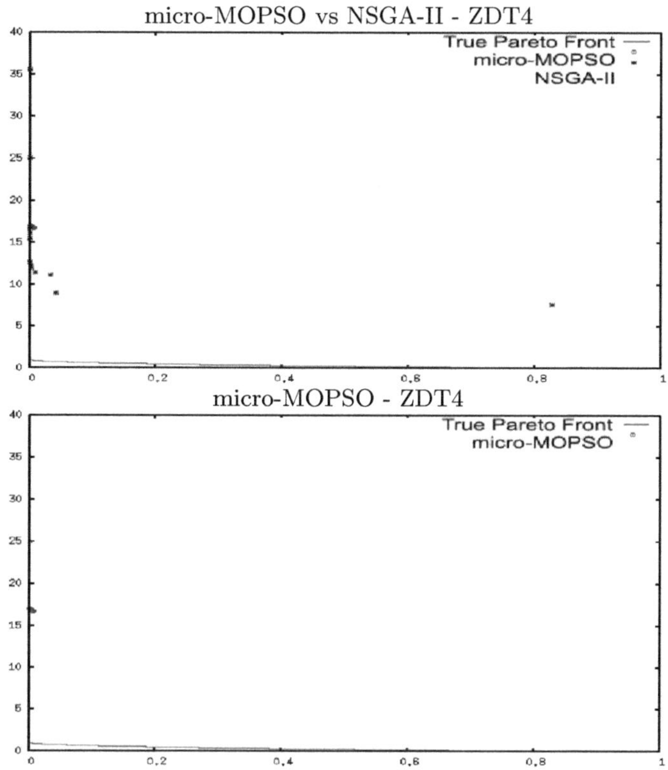

Fig. 4.5. Graphical comparison of the Pareto fronts found by our micro-MOPSO (right) and the NSGA-II (left) for ZDT4.

4.6.2 Experiments

For assessing the performance of the micro-MOPSO, we used five test functions from the ZDT (Zitzler-Deb-Thiele) benchmark described in [39]. Additionally, we adopted Kursawe's test function [17] and Viennet's test funtion [37]. These test functions contain characteristics that make them difficult to solve using MOEAs.

We performed thirty independent runs for each test function and we compared our results with respect to the NSGA-II [8]. The number of objective function evaluations performed was set to 3000 in all cases.

For obtaining the results reported next, we adopted the following parameters, which were obtained after numerous experiments:

- w = random number from a uniform distribution in the range [0,1]
- $C_1 = C_2 = 1.8$
- population size = 5 particles

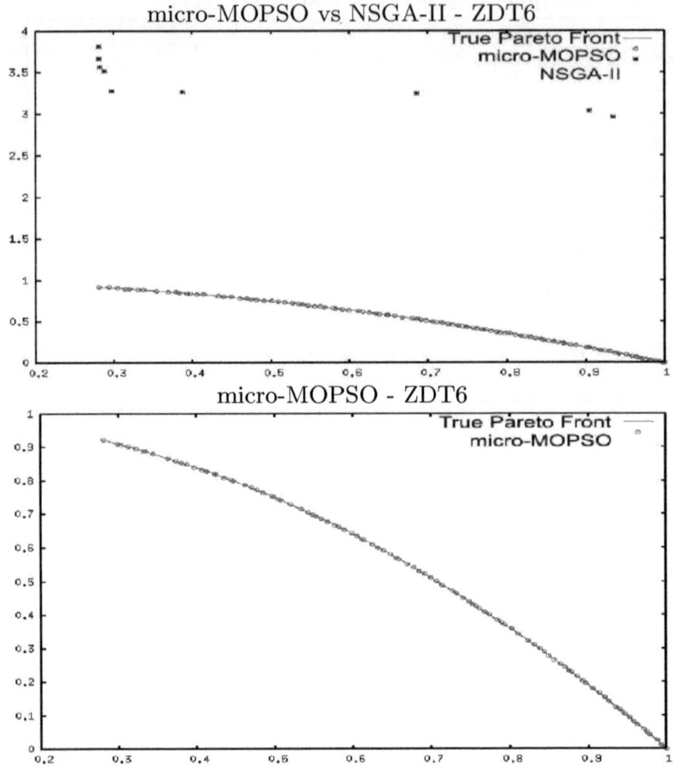

Fig. 4.6. Graphical comparison of the Pareto fronts found by our micro-MOPSO (right) and the NSGA-II (left) for ZDT6.

- number of generations = 600
- final archive bound (FAB) = 100
- auxiliary archive bound (AAB)= 200
- population percent for leader selection (pps) = 0.20
- number of replacement generations = 100
- number of replacement particles = 2
- mutation rate = 0.1

4.6.3 Results

The summary of the results obtained are shown in Table 4.1. This table shows the mean values and the standard deviations for each of the three performance measures adopted over the 30 independent runs performed.

The results show that our micro-MOPSO produced the best mean values in almost all cases. The graphical results shown in Figures 4.2 to 4.8, clearly reflect that our proposed micro-MOPSO outperforms the NSGA-II. In test functions ZDT1, ZDT2, ZDT3 and ZDT6, we can see that the NSGA-II is

Table 4.1. Comparison of results between our micro-MOPSO and the NSGA-II. σ refers to the standard deviation over the 30 runs performed.

	ER				DGI				S			
	micro-MOPSO		NSGA-II		micro-MOPSO		NSGA-II		micro-MOPSO		NSGA-II	
	Mean	σ	Mean	σ	Mean	σ	Mean	σ	Mean	σ	Mean	σ
ZDT1	**0.53**	0.4796	1.0	0.0	**0.0003**	0.0002	0.0151	0.002	**0.005**	0.0004	0.031	0.009
ZDT2	**0.496**	0.334	1.0	0.0	**0.0009**	0.0036	0.0339	0.006	**0.0048**	0.0005	0.048	0.0182
ZDT3	**0.935**	0.244	1.0	0.0	**0.002**	0.0043	0.0216	0.002	**0.0085**	0.009	0.034	0.009
ZDT4	1.0	0.0	1.0	0.0	1.585	0.790	**0.594**	0.1782	**0.0001**	0.0004	5.321	3.837
ZDT6	**0.133**	0.001	1.0	0.0	**0.00006**	0.0	0.0570	0.004	**0.033**	0.149	0.124	0.085
Kur	0.954	0.392	**0.837**	0.093	**0.003**	0.0007	0.004	0.0005	**0.082**	0.028	0.10	0.025
Vnt	**0.592**	0.44	0.6	0.36	**0.0006**	0.0005	0.002	0.004	**0.04**	0.02	0.061	0.010

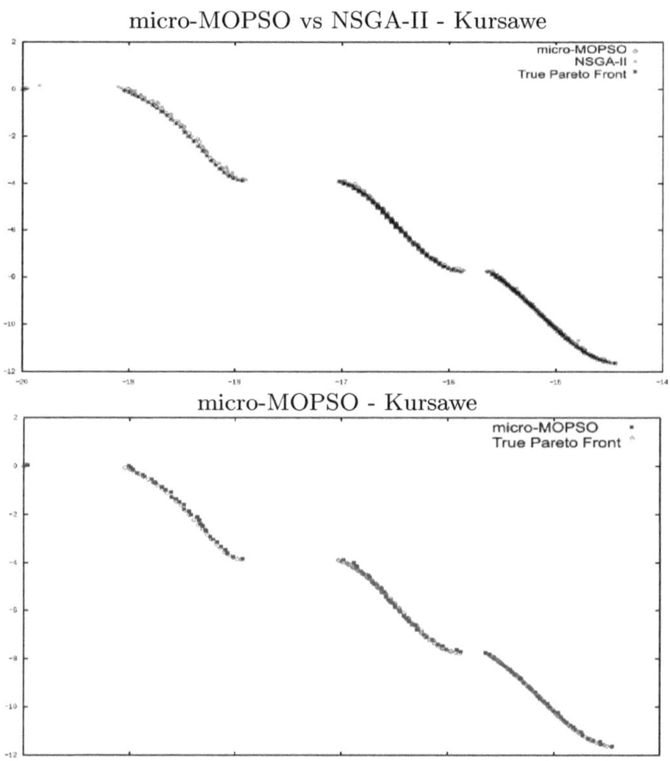

Fig. 4.7. Graphical comparison of the Pareto fronts found by our micro-MOPSO (right) and the NSGA-II (left) for Kursawe's test function.

far away from the true Pareto front, whereas our micro-MOPSO has already converged to the true Pareto front after performing only 3000 fitness function evaluations. Figures 4.7 and 4.8 show that our micro-MOPSO can cover most of the Pareto front even when it is discontinuous (ZDT3 and Kursawe). The

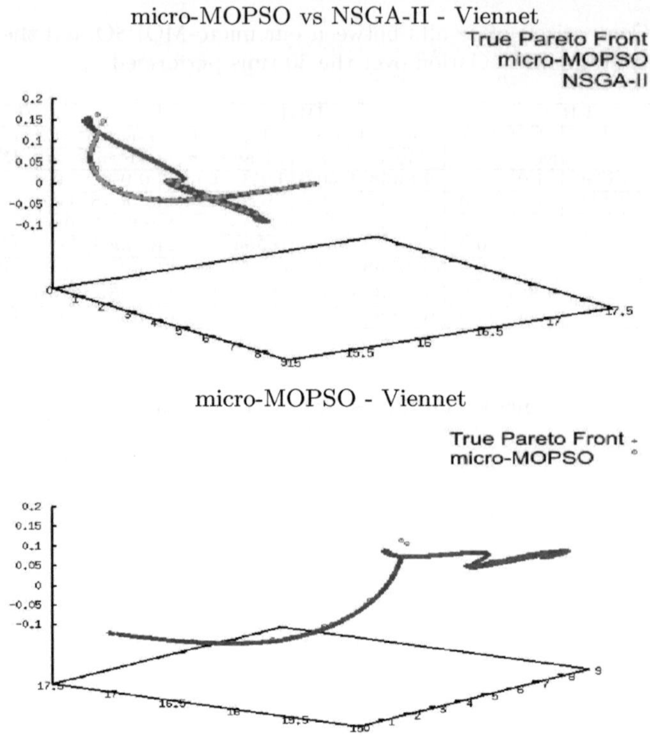

Fig. 4.8. Graphical comparison of the Pareto fronts found by our micro-MOPSO (right) and the NSGA-II (left) for Viennet's test function.

results and the figures show that the spread of solutions of our micro-MOPSO is evidently good enough for almost all the test functions (except for ZDT4 in which both algorithms have a poor performance) in spite of the low number of objective function evaluations performed.

4.7 Conclusions and Future Work

We have proposed the use of a PSO algorithm with a very small population size (only five particles) for solving unconstrained multi-objective optimization problems. The proposed approach first selects the leader and then selects the neighborhood for constructing the swarm around the leader. It also uses a mutation operator, a reinitialization process, and a mechanism based on the crowding distance for selecting leaders. Our proposed approach was able to provide a better spread of the solutions obtained, as well as a faster convergence, when compared to the NSGA-II, in several test functions, in which only 3000 objective functions were performed. These results are very encouraging and indicate that our proposed approach could be a viable alternative for solving problems in which the evaluations of the objective functions are very expensive (computationally speaking).

As part of our future work, we are interested in incorporating a constraint-handling mechanism into our approach. We are also interested in adopting clustering techniques in order to provide a better spread of solutions. Additionally, we also want to study the sensitivity of our micro-MOPSO to its parameters, with the aim of finding a set of parameters (or a self-adaptation mechanism) that allows us to improve the performance and the robustness of our approach. Finally, we are also interested in experimenting with other types of reinitialization processes, since they could improve the convergence rate of our algorithm (i.e., we could reduce the number of objective function evaluations performed) as well as the quality of the results achieved.

Acknowledgements

The second author acknowledges support from CONACyT project no. 45683-Y.

References

1. Alvarez-Benitez, J.E., Everson, R.M., Fieldsend, J.E.: A MOPSO algorithm based exclusively on pareto dominance concepts. In: Coello, C.A.C., Hernández Aguirre, A., Zitzler, E. (eds.) EMO 2005. LNCS, vol. 3410, pp. 459–473. Springer, Heidelberg (2005)
2. Andrews., P.S.: An investigation into mutation operators for particle swarm optimization. In: Proceedings of the 2006 IEEE Congress on Evolutionary Computation (CEC 2006), Vancouver, Canada, July 2006, pp. 3789–3796 (2006)
3. Bartz-Beielstein, T., Limbourg, P., Parsopoulos, K.E., Vrahatis, M.N., Mehnen, J., Schmitt, K.: Particle Swarm Optimizers for Pareto Optimization with Enhanced Archiving Techniques. In: Proceedings of the 2003 Congress on Evolutionary Computation (CEC 2003), Canberra, Australia, December 2003, vol. 3, pp. 1780–1787. IEEE Press, Los Alamitos (2003)
4. Coello, C.A.C., Pulido, G.T.: Multiobjective optimization using a micro-genetic algorithm. In: Spector, L., Goodman, E.D., Wu, A., Langdon, W., Voigt, H.M., Gen, M., Sen, S., Dorigo, M., Pezeshk, S., Garzon, M., Burke, E. (eds.) Genetic and Evolutionary Computation Conference, GECCO,2001, pp. 274–282. Morgan Kaufmann Publishers, San Francisco (2001)
5. Coello, C.A.C., Pulido, G.T.: A Micro-Genetic Algorithm for Multiobjective Optimization. In: Zitzler, E., Deb, K., Thiele, L., Coello, C.A.C., Corne, D.W. (eds.) EMO 2001. LNCS, vol. 1993, pp. 126–140. Springer, Heidelberg (2001)
6. Coello, C.A.C., Pulido, G.T., Lechuga, M.S.: Handling Multiple Objectives With Particle Swarm Optimization. IEEE Transactions on Evolutionary Computation 8(3), 256–279 (2004)
7. Coello, C.A.C., Van Veldhuizen, D.A., Lamont, G.B.: Evolutionary Algorithms for Solving Multi-Objective Problems, 2nd edn. Springer, New York (2007)

8. Deb, K., Pratap, A., Agarwal, S., Meyarivan, T.: A Fast and Elitist Multiobjective Genetic Algorithm: NSGA–II. IEEE Transactions on Evolutionary Computation 6(2), 182–197 (2002)
9. Eberhart, R., Kennedy, J.: Particle swarm optimization. In: Proceedings of the IEEE International Conference on Neural Networks, pp. 1942–1948. IEEE Press, Los Alamitos (1995)
10. Engelbrecht., A.P.: Fundamentals of Computational Swarm Intelligence. John Wiley & Sons Ltd., England (2005)
11. Esquivel, S.C., Coello, C.A.C.: On the use of particle swarm optimization with multimodal functions. In: Proceedings of the 2003 IEEE Congress on Evolutionary Computation (CEC 2003), Canberra, Australia, pp. 1130–1136. IEEE Press, Los Alamitos (2003)
12. Fieldsend, J.E., Singh, S.: A Multi-Objective Algorithm based upon Particle Swarm Optimisation, an Efficient Data Structure and Turbulence. In: Proceedings of the 2002 U.K. Workshop on Computational Intelligence, Birmingham, UK, September 2002, pp. 37–44 (2002)
13. Fuentes Cabrera, J.C., Coello, C.A.C.: Handling constraints in particle swarm optimization using a small population size. In: Gelbukh, A., Kuri Morales, Á.F. (eds.) MICAI 2007. LNCS, vol. 4827, pp. 41–51. Springer, Heidelberg (2007)
14. Kennedy, J., Eberhart, R.C.: Swarm Intelligence. Morgan Kauffmann Publishers, San Francisco (2001)
15. Knowles, J., Corne, D.: The pareto archived evolution strategy: A new baseline algorithm for pareto multiobjective optimisation. In: Angeline, P.J., Michalewicz, Z., Schoenauer, M., Yao, X., Zalzala, A. (eds.) Proceedings of the Congress on Evolutionary Computation, Mayflower Hotel, Washington D.C, vol. 1, pp. 98–105. IEEE Press, Los Alamitos (1999)
16. Krishnakumar, K.: Micro-genetic algorithms for stationary and non-stationary function optimization. In: SPIE Proceedings: Intelligent Control and Adaptive Systems, vol. 1196, pp. 289–296 (1989)
17. Kursawe, F.: A Variant of Evolution Strategies for Vector Optimization. In: Schwefel, H.-P., Männer, R. (eds.) PPSN 1990. LNCS, vol. 496, pp. 193–197. Springer, Heidelberg (1991)
18. Laumanns, M., Thiele, L., Deb, K., Zitzler, E.: Combining Convergence and Diversity in Evolutionary Multi-objective Optimization. Evolutionary Computation 10(3), 263–282 (2002)
19. Li, X.: A Non-dominated Sorting Particle Swarm Optimizer for Multiobjective Optimization. In: Cantú-Paz, E., Foster, J.A., Deb, K., Davis, L., Roy, R., O'Reilly, U.-M., Beyer, H.-G., Kendall, G., Wilson, S.W., Harman, M., Wegener, J., Dasgupta, D., Potter, M.A., Schultz, A., Dowsland, K.A., Jonoska, N., Miller, J., Standish, R.K. (eds.) GECCO 2003. LNCS, vol. 2723, pp. 37–48. Springer, Heidelberg (2003)
20. Michalewicz, Z.: Genetic Algorithms + Data Structures = Evolution Programs. Springer, Heidelberg (1996)
21. Miettinen, K.M.: Nonlinear Multiobjective Optimization. Kluwer Academic Publishers, Boston (1999)
22. Moore, J., Chapman, R.: Application of particle swarm to multiobjective optimization (1999)

23. Mostaghim, S., Teich, J.: The Role of ε-dominance in Multi Objective Particle Swarm Optimization Methods. In: Proceedings of the 2003 Congress on Evolutionary Computation (CEC 2003), Canberra, Australia, December 2003, vol. 3, pp. 1764–1771. IEEE Press, Los Alamitos (2003)
24. Mostaghim, S., Teich, J.: Strategies for Finding Good Local Guides in Multi-objective Particle Swarm Optimization (MOPSO). In: 2003 IEEE Swarm Intelligence Symposium Proceedings, Indianapolis, Indiana, USA, April 2003, pp. 26–33. IEEE Service Center, Los Alamitos (2003)
25. Mostaghim, S., Teich, J.: Covering Pareto-optimal Fronts by Subswarms in Multi-objective Particle Swarm Optimization. In: 2004 Congress on Evolutionary Computation (CEC 2004), Portland, Oregon, USA, June 2004, vol. 2, pp. 1404–1411. IEEE Service Center, Los Alamitos (2004)
26. Sierra, M.R., Coello, C.A.C.: Improving PSO-Based Multi-objective Optimization Using Crowding, Mutation and ϵ-Dominance. In: Coello, C.A.C., Hernández Aguirre, A., Zitzler, E. (eds.) EMO 2005. LNCS, vol. 3410, pp. 505–519. Springer, Heidelberg (2005)
27. Reyes-Sierra, M., Coello, C.A.C.: Multi-Objective Particle Swarm Optimizers: A Survey of the State-of-the-Art. International Journal of Computational Intelligence Research 2(3), 287–308 (2006)
28. Sierra, M.R., Coello, C.A.C.: A Study of Techniques to Improve the Efficiency of a Multi-Objective Particle Swarm Optimizer. In: Yang, Y.S., Ong, Y. (eds.) Evolutionary Computation in Dynamic and Uncertain Environments, pp. 269–296. Springer, Heidelberg (2007) ISBN 978-3-540-49772-1
29. Schoeman, I., Engelbrecht, A.: Niching for Dynamic Environments using Particle Swarm Optimization. In: Wang, T.-D., Li, X., Chen, S.-H., Wang, X., Abbass, H.A., Iba, H., Chen, G.-L., Yao, X. (eds.) SEAL 2006. LNCS, vol. 4247, pp. 134–141. Springer, Heidelberg (2006)
30. Schott, J.R.: Fault Tolerant Design Using Single and Multicriteria Genetic Algorithm Optimization. Master's thesis, Department of Aeronautics and Astronautics, Massachusetts Institute of Technology, Cambridge, Massachusetts (May 1995)
31. Shi, Y., Eberhart, R.C.: A modified particle swarm optimizer. In: Proceedings of the 1998 IEEE Congress on Evolutionary Computation, pp. 69–73. IEEE Press, Los Alamitos (1998)
32. Tan, C.H., Goh, C.K., Tan, K.C., Tay, A.: A Cooperative Coevolutionary Algorithm for Multiobjective Particle Swarm Optimization. In: 2007 IEEE Congress on Evolutionary Computation (CEC 2007), Singapore, September 2007, pp. 3180–3186. IEEE Press, Los Alamitos (2007)
33. Toscano Pulido, G., Coello, C.A.C.: The Micro Genetic Algorithm 2: Towards Online Adaptation in Evolutionary Multiobjective Optimization. In: Fonseca, C.M., Fleming, P.J., Zitzler, E., Deb, K., Thiele, L. (eds.) EMO 2003. LNCS, vol. 2632, pp. 252–266. Springer, Heidelberg (2003)
34. Toscano Pulido, G., Coello, C.A.C.: Using clustering techniques to improve the performance of a multi-objective particle swarm optimizer. In: Deb, K., et al. (eds.) GECCO 2004. LNCS, vol. 3102, pp. 225–237. Springer, Heidelberg (2004)
35. Tripathi, P.K., Bandyopadhyay, S., Pal, S.K.: Multi-objective particle swarm optimization with time variant inertia and acceleration coefficients. Information Sciences 177(22), 5033–5049 (2007)

36. Van Veldhuizen, D.A., Lamont, G.B.: Multiobjective Evolutionary Algorithm Test Suites. In: Carroll, J., Haddad, H., Oppenheim, D., Bryant, B., Lamont, G.B. (eds.) Proceedings of the 1999 ACM Symposium on Applied Computing, San Antonio, Texas, pp. 351–357. ACM, New York (1999)
37. Viennet, R., Fontiex, C., Marc, I.: Multicriteria Optimization Using a Genetic Algorithm for Determining a Pareto Set. International Journal of Systems Science 27(2), 255–260 (1996)
38. Xiao-hua, Z., Hong-yun, M., Li-cheng, J.: Intelligent Particle Swarm Optimization in Multiobjective Optimization. In: 2005 Congress on Evolutionary Computation, Edinburgh, Scotland, UK, September 2005, pp. 714–719. IEEE Press, Los Alamitos (2005)
39. Zitzler, E., Deb, K., Thiele, L.: Comparison of multiobjective evolutionary algorithms: Empirical results. Evolutionary Computation 8(2), 173–195 (2000)

5

Dynamic Multi-objective Optimisation Using PSO

Mardé Greeff[1,2] and Andries P. Engelbrecht[2]

[1] CSIR, Meraka Insitute, P.O. Box 395, Pretoria, 0001, South Africa
mgreeff@csir.co.za
[2] University of Pretoria, Faculty of Engineering,
Built Environment and Information Technology,
Department of Computer Science, Pretoria, 0001, South Africa
engel@cs.up.ac.za

Optimisation problems occur in many situations and aspects of modern life. In reality, many of these problems are dynamic in nature, where changes can occur in the environment that influence the solutions of the optimisation problem. Many methods use a weighted average approach to the multiple objectives. However, generally a dynamic multi-objective optimisation problem (DMOOP) does not have a single solution. In many cases the objectives (or goals) are in conflict with one another, where an improvement in one objective leads to a worse solution for at least one of the other objectives. The set of solutions that can be found where no other solution is better for all the objectives, is called the Pareto optimal front (POF) and the solutions are called non-dominated solutions. The goal when solving a DMOOP is not to find a single solution, but to find the POF. This chapter introduces the usage of the vector evaluated particle swarm optimiser (VEPSO) to solve DMOOPs. Every objective is solved by one swarm and the swarms share knowledge amongst each other about the objective that it is solving. Not much work has been done on using this approach in dynamic environments. This chapter discusses this approach, as well as the effect that various ways of transferring knowledge between the swarms, together with the population size and various response methods to a detected change, have on the performance of the algorithm.

5.1 Introduction

In the world of today optimisation problems occur in a vast variety of situations and aspects of modern life. However, in reality many of these problems are dynamic in nature, where changes can occur in the environment that

N. Nedjah et al. (Eds.): Multi-Objective Swarm Intelligent Systems, SCI 261, pp. 105–123.
springerlink.com

influence the solutions of the optimisation problem. Examples of dynamic optimisation problems can be found in a vast range of scheduling problems, such as timetables, air traffic control, routing in telecommunication networks and target tracking in military operations. Most optimisation problems have more than one objective, e.g. the goals for a manufacturing process might be to maximise the number of products that are manufactured, minimise the time required to manufacture a specific number of products, minimise the time that any machine is idle and minimising the cost. Using a specific machine can be more expensive than another, however the more expensive machine might require less time to manufacture the same number of products than a machine that is cheaper to operate. If you want to manufacture the maximum number of products, using the more expensive machine will minimise the time required, but will increase the cost. Optimisation problems that have more than one objective are called multi-objective optimisation (MOO) problems. Normally these objectives are in conflict with one another, i.e. a better solution for one objective leads to a worse solution for at least one of the other objectives. The set of solutions that can be found where no other solution is better for all the objectives, is called the Pareto optimal front (POF), and the solutions are called non-dominated solutions.

When changes occur in the environment that influence the solutions of the MOO problem, the goal becomes to track the changing POF. In order to achieve this, the algorithm should be able to firstly detect that a change has occcurred, and then respond to the change in an appropriate way. If one or more machines used in a manufacturing process (such as the example explained above) break down, it will cause a change in the required solutions and will therefore be an example of a dynamic MOO problem (DMOOP). Comparing one algorithm's performance against another when solving a DMOOP is not a trivial task, since in many real world problems the true POF is unknown. Therefore, benchmark functions with different POF characteristics are used to test the efficiency of an algorithm. To measure the performance of an algorithm, peformance metrics are used.

Greeff and Engelbrecht proposed using the Vector Evaluated Particle Swarm Optimisation (VEPSO) algorithm to solve DMOOPs [1]. When the VEPSO approach is used, each swarm solves only one objective function and then the swarms share knowledge amongst each other. In this chapter the performance of the VEPSO algorithm, using different ways to transfer knowledge between the swarms, are discussed. The effect of the way in which knowledge is transferred between the swarms, in combination with swarm sizes and the effect of various responses to detected changes in the environment, on the performance of the VEPSO algorithm, is also discussed in this chapter.

The rest of the chapter's layout is as follows: Section 5.2 provides background information and highlights related work that is relevant for the research discussed in this paper. Section 5.3 provides an overview of the VEPSO algorithm and the changes that has been made to the algorithm for DMOOPs. The experiments that have been conducted are discussed in Sect. 5.4. The benchmark functions and performance metrics that have been used to test the algorithm's performance, are highlighted in Sects. 5.4.1 and 5.4.2 respectively. The statistical methods that were used for analyses are presented in Sect. 5.4.3. Section 5.5 describes the results of the experiments. Finally, Sect. 5.6 provides a summary and conclusions on the work presented in this chapter.

5.2 Background

Eberhart and Kennedy [2] introduced Particle Swarm Optimisation (PSO), a population-based optimisation method inspired by the social behaviour of bird flocks. Each PSO swarm consists of a number of particles that searches solutions by moving through the search space. Each individual particle has a current position, $\mathbf{x}_i$, velocity, $\mathbf{v}_i$, and personal best position, $\mathbf{y}_i$, where the particle had the smallest error with regards to the objective function. The position amongst all the particles personal best positions that resulted in the smallest error, is called the global best position, denoted as $\hat{\mathbf{y}}$. During each iteration every particle's new position is determined by adding the new velocity to the particle's current position.

Dynamic single-objective optimization problems have successfully been solved using PSO ([3, 4, 5, 6]). When dealing with dynamic problems, where a change in the environment results in a change in the solutions, it is vital that the algorithm can detect that a change has occurred. The concept of a sentry particle has been introduced by Carlisle and Dozier [7]. When using the sentry particle approach to detect whether a change has occurred in the environment, a random number of sentry particles are selected after each iteration. These particles are then re-evaluated before the next iteration, where each particle's current fitness value is compared with its previous fitness value, i.e. its fitness value after the previous iteration. If these two values differ more than a specified value, the swarm is alerted that a change has occurred. Hu and Eberhart [8] suggested that the global best, and global second best particles should be used as sentry particles.

If a change has been detected, the algorithm should respond appropriately to the change. One approach, suggested by Carlise [9], is to re-calculate each particle's personal best. If the new personal best value is less fit than the particles's current position, its personal best value is replaced by its current position. This comparison enables retaining valid past experience. Another approach is to re-initialize a percentage of the swarm's population, preventing that the swarm remains in a small area after the change has occurred, and enabling a portion of the particles to retain their memory, which could be valuable information if the change is small. Cui et al. [6] proposed the usage of an evaporation constant, with a value between zero and one, that is used to update the particle's best fitness value. The particle's memory will gradually decrease over time proportially to the evaporation constant. At a certain point in time the particle's current fitness will be better than the decreased fitness value. When this happens, the decreased fitness value will be replaced by the particle's current fitness. With the evaporation constant approach, the environment is not monitored by any particles, as is the case with the usage of sentry particles.

Much work has been done on using PSO for multi-objective optimisation (MOO) problems. Reyes-Sierra and Coello Coello provides a comprehensive review of the various PSO approaches that were used to solve MOO problems [10]. Recently evolutionary algorithms (EAs) ([11, 12, 13]) and PSO ([14, 15]) have been used to solve DMOOPs. Since changes occur in the environment, the goal is to track the changing POF. In order to test and analyse an algorithm's capability of tracking a dynamic POF, benchmark functions are used. Jin and Sendhof [16] introduced how to define a dynamic two-objective optimisation problem by reformulating a three-objective optimisation test function. Another approach where dynamic problems are created by replacing objectives with new ones at specific times, were presented by Guan et al. [11]. Other benchmark functions were developed by adapting static MOO benchmark functions to dynamic ones. A number of test functions for DMOO based on the static MOO two-objective ZDT functions [17] and the scalable DTLZ functions [18] were developed by Farina et al. [19]. Some adaptions to these test functions were proposed in ([12, 20]). Mehnen et al. [12] proposed DSW functions that are adapted from the static MOO function problem of Schaffer [21] and others added noise to Deb's functions ([22, 23]).

Once the benchmark functions were used to test an algorithm's capability of tracking a dynamic POF, the data has to be analysed and the tracking capability of various algorithms should be compared against one another. In order to compare one algorithm's performance against another algorithm, performance metrics are required. For DMOOP two groups of performance metrics exist, namely performance metrics where the true POF is known and performance metrics where the true POF is unknown. The convergence, measured by the generational distance proposed by Van Veldhuizen [24], and spread or distribution of the solutions are often used to measure an algorithm's performance when the POF is known ([23, 25]). The reversed generational distance and the collective mean error were proposed as performance metrics by Branke et al. [14]. Another metric, the $HVR(t)$ metric, represents the ratio of the hypervolume of the solutions and the hypervolume of the known POF at a specific time ([24, 14]). Li et al. [14] proposed a metric of spacing that can be used when the true POF is unknown. Measures of accuracy, stability and reaction capacity of an algorithm, that are based on the calculation of the hypervolume of the non-dominated solution set, was proposed by Cámara et al. [15].

The next section discusses the Vector Evaluated Particle Swarm Optimisation approach that is used to solve MOO problems.

5.3 Vector Evaluated Particle Swarm Optimisation

The Vector Evaluated Particle Swarm Optimisation (VEPSO), a multi-swarm variation of PSO and inspired by the Vector Evaluated Genetic Algorithm (VEGA) [26], was introduced by Parsopoulos et al. [27].

With the VEPSO approach each swarm solves only one objective function and shares its knowledge with the other swarms. The shared knowledge is then used to update the velocity of the particles as indicated in Eqs. (5.1) and (5.2) below:

$$v_i^j(t+1) = w^j v_i^j(t) + c_1^j \mathbf{r_1}(y_i^j(t) - x_i^j(t)) + c_2^j \mathbf{r_2}(\hat{y}_i^s(t) - x_i^j(t)) \tag{5.1}$$

$$x_i^j(t+1) = x_i^j(t) + v_i^j(t+1) \tag{5.2}$$

where n represents the dimension with $i = 1, \ldots, n$; m represents the number of swarms with $j = 1, \ldots, m$ as the swarm index; $\hat{\mathbf{y}}^s$ is the global best of the s-th swarm; c_1^j and c_2^j are the cognitive and social parameters of the j-th swarm respectively; $\mathbf{r_1}, \mathbf{r_2} \in [0,1]^n$; w^j is the inertia weight of the j-th swarm; and $s \in [1, \ldots, j-1, j+1, \ldots, m]$ represents the index of a respective swarm. The index s can be set up in various ways, affecting the topology of the swarms in VEPSO.

The VEPSO approach for MOO problems is presented in Algorithm 1. The set of solutions found so far, forming the found POF, are stored in an archive. If the archive is full, and the new solutions are non-dominated and do not already exist in the archive, solutions are selected for removal from the archive to make space for the new solutions based on the distance metric [28].

Algorithm 1. VEPSO for MOO problems

```
for number of iterations do
    perform vepso iteration
    if new solutions are non-dominated and don't exist in archive then
        if space in archive then
            add new solutions to archive
        else
            remove solutions from archive
            add new solutions to archive
        end if
    end if
end for
```

In order to solve DMOO problems the algorithm has to be adapted. The changes that were made to the VEPSO approach to solve DMOO problems are discussed next.

5.3.1 VEPSO for Dynamic Multi-objective Optimisation

The VEPSO approach discussed above is used to solve MOO problems. In order to use this approach for DMOO problems, the algorithm must be able to detect that a change has occurred and then respond to the change in an appropriate manner. Sentry particles are used to detect a change (as discussed in Sect. 5.2). When a change is detected, a percentage of the swarm's particles are re-initialised. Re-initialisation of a particle entails re-setting its position and then re-evaluating its personal best and the neighbourhood best. The VEPSO approach for DMOO problems is presented in Algorithm 2. The next Section discusses the experiments that were performed to evaluate the ability of this approach to solve DMOO problems.

Algorithm 2. VEPSO for DMOO problems

```
for number of iterations do
    check whether a change has occurred
    if change has occurred
        respond to change
        re-evaluate solutions in archive
        remove dominated solutions from archive
    perform vepso iteration
    if new solutions are non-dominated and don't exist in archive
        if space in archive
            add new solutions to archive
        else
            remove solutions from archive
            add new solutions to archive
    select sentry particles
```

5.4 Experiments

This section describes the experiments that were conducted to test the performance of VEPSO when solving DMOOPs. The benchmark functions and performanc metrics, discussed in Sects. 5.4.1 and 5.4.2 respectively, were used to test the performance of variations of the VEPSO algorithm presented in Sect. 5.3.1.

To test the effect of various ways of transferring knowledge between the swarms, a ring topology is used, as well as a random selection, i.e. the swarm whose knowledge is used is selected randomly. These different topologies are tested using a varied population size, where the swarms' number of particles are varied between 10, 20, 30 and 40 particles respectively. Furthermore, various responses to change are used, where either 10%, 20% or 30% of the swarm's population is re-initialised. Re-initialisation of particles was either done for all swarms, or only for the swarm that is solving the objective function that has changed.

All experiments consisted of 30 runs and each run consisted of 1 000 iterations. The frequency of change, τ_t, was set to 5 as suggested by Farina et al. [19], i.e. during each run the functions change every 5 iterations, resulting in 200 changes per run. The PSO parameters were set to values that lead to convergent behaviour [29], namely w=0.72 and $c1 = c2 = 1.49$.

5.4.1 Benchmark Functions

To test the performance of VEPSO solving DMOOPs with various types of POFs, with two or more objective functions, four functions proposed by Farina et al. [19] are used. Below, τ is the generation counter, τ_t is the number of iterations for which t remains fixed and n_t is the number of distinct steps in t.

$$\text{FDA1}=\begin{cases} Minimize: f(\mathbf{x},t)=(f_1(\mathsf{x}_\mathsf{I},t), g(\mathbf{x}_{\mathbf{II}},t)\cdot h(\mathbf{x}_{\mathbf{III}}, f_i(\mathsf{x}_\mathsf{I},t), g(\mathbf{x}_{\mathbf{II}},t),t)) \\ f_1(\mathsf{x}_\mathsf{I}) = x_i \\ g(\mathbf{x}_{\mathbf{II}}) = 1+\sum_{x_i\in\mathbf{x}_{\mathbf{II}}}(x_i - G(t))^2 \\ h(f_1,g) = 1-\sqrt{\frac{f_1}{g}} \\ where: \\ G(t) = \sin(0.5\pi t),\quad t = \frac{1}{n_t}\left\lfloor\frac{\tau}{\tau_t}\right\rfloor \\ \mathsf{x}_\mathsf{I}\in[0,1];\quad \mathbf{x}_{\mathbf{II}} = (x_2,\ldots,x_n)\in[-1,1] \end{cases} \tag{5.3}$$

The function parameters were set to $n = 20$ and $n_t = 10$ (as suggested by [19]). Function FDA1 (Eq. (5.3)) has a convex POF where the values in the decision variable space changes, but the values in the objective space remains the same.

$$\text{FDA2}=\begin{cases} Minimize: f(\mathsf{x},t)=(f_1(\mathbf{x}_{\mathbf{I}},t), g(\mathbf{x}_{\mathbf{II}},t)\cdot h(\mathbf{x}_{\mathbf{III}}, f_i(\mathbf{x}_{\mathbf{I}},t), g(\mathbf{x}_{\mathbf{II}},t),t)) \\ f_1(\mathbf{x}_{\mathbf{I}}) = x_i \\ g(\mathbf{x}_{\mathbf{II}}) = 1+\sum_{x_i\in\mathbf{x}_{\mathbf{II}}} x_i^2 \\ h(f_1,g) = 1-\frac{f_1}{g}^{(H(t)+\sum_{x_i\in\mathbf{x}_{\mathbf{III}}}(x_i-H(t))^2)} \\ where: \\ H(t) = 0.75+0.75\sin(0.5\pi t), t = \frac{1}{n_t}\left\lfloor\frac{\tau}{\tau_t}\right\rfloor \\ \mathbf{x}_{\mathbf{I}}\in[0,1]; \mathbf{x}_{\mathbf{II}},\mathbf{x}_{\mathbf{III}}\in[-1,1] \end{cases} \tag{5.4}$$

For FDA2 the parameters were set to the following: $|\mathbf{x}_{\mathbf{II}}| = |\mathbf{x}_{\mathbf{III}}| = 15$ and $n_t = 10$ (as suggested by [19]). Function FDA2 (Eq. (5.4)) has a POF that changes from a convex to a non-convex shape, while the values in the decision variable space remains the same.

$$\mathsf{FDA4} = \begin{cases} Minimize: f_1(\mathbf{x}), \ldots, f_m(\mathbf{x}) \\ min_x : f_1(\mathbf{x}) = (1 + g(\mathbf{x_{II}})) \prod_{i=1}^{m-1} \cos(\frac{x_i \pi}{2}) \\ min_x : f_k(\mathbf{x}) = (1 + g(\mathbf{x_{II}}))(\prod_{i=1}^{m-k} \cos(\frac{x_i \pi}{2})) \sin(\frac{x_{m-k+1}\pi}{2}) \\ min_x : f_m(\mathbf{x}) = (1 + g(\mathbf{x_{II}})) \sin(\frac{x_1 \pi}{2}) \\ where: \\ g(\mathbf{x_{II}}) = \sum_{x_i \in \mathbf{x_{II}}} (x_i - G(t))^2 \\ G(t) = |sin(0.5\pi t)| \\ F(t) = 10^{2\sin(0.5\pi t)}, \;\; t = \frac{1}{n_t} \left\lfloor \frac{\tau}{\tau_t} \right\rfloor \\ \mathbf{x_{II}} = (x_m, \ldots, x_n); \; x_i \in [0,1], \forall i = 1, \ldots, n; \; k = 2, \ldots, m-1 \end{cases} \tag{5.5}$$

For FDA4 the parameters were set as $n = m + 9$, $|\mathbf{x_{II}}| = 10$, $n_t = 10$ (as suggested by [19]) and $m = 3$ for a 3-objective function. Function FDA4 (Eq. (5.5)) has a non-convex shaped POF where the values in the decision variable space changes, but the values in the objective space remains the same.

$$\mathsf{FDA5} = \begin{cases} Minimize: f_1(\mathbf{x}), \ldots, f_m(\mathbf{x}) \\ min_x : f_1(\mathbf{x}) = (1 + g(\mathbf{x_{II}})) \prod_{i=1}^{m-1} \cos(\frac{y_i \pi}{2}) \\ min_x : f_k(\mathbf{x}) = (1 + g(\mathbf{x_{II}}))(\prod_{i=1}^{m-k} \cos(\frac{y_i \pi}{2})) \sin(\frac{y_{m-k+1}\pi}{2}) \\ min_x : f_m(\mathbf{x}) = (1 + g(\mathbf{x_{II}})) \sin(\frac{y_1 \pi}{2}) v \\ where: \\ g(\mathbf{x_{II}}) = G(t) + \sum_{x_i \in \mathbf{x_{II}}} (x_i - G(t))^2 \\ y_i = x_i^{F(t)}, \forall i \in [1, \ldots, (m-1)] \\ G(t) = |sin(0.5\pi t)| \\ F(t) = 1 + 100\sin^4(0.5\pi t), \;\; t = \frac{1}{n_t} \left\lfloor \frac{\tau}{\tau_t} \right\rfloor \\ \mathbf{x_{II}} = (x_m, \ldots, x_n); \; x_i \in [0,1], \forall i = 1, \ldots, n; \; k = 2, \ldots, m-1 \end{cases} \tag{5.6}$$

Identical function parameters and values as those specified for FDA4 were used here. Function FDA5 (Eq. (5.6)) has a non-convex shaped POF where the values in both the decision variable space and the objective space changes.

5.4.2 Performance Metrics

This chapter assumes that the POF of the benchmark functions are unknown, since in reality this will often be the case. The performance metrics that are used to compare the performance of the VEPSO variations are discussed below.

Spacing

The metric of spacing [23] indicates how evenly the non-dominated solutions are distributed along the discovered POF, and is defined as

$$S = \frac{1}{n_{PF}} \left[\frac{1}{n_{PF}} \sum_{i=1}^{n_{PF}} (d_i - \overline{d})^2 \right]^{\frac{1}{2}}, \overline{d} = \frac{1}{n_{PF}} \sum_{i=1}^{n_{PF}} d_i; \quad \overline{S} = \frac{1}{n_r} \sum_{j=1}^{n_r} S_j^i \tag{5.7}$$

where n_{PF} is the number of non-dominated solutions found and d_i is the euclidean distance, in the objective space, between a non-dominated solution i and its nearest non-dominated solution.

To compare one algorithm against another, the average spacing metric $\overline{S}$ is calculated for each iteration just before a change in the environment occurs.

Hypervolume

The S-metric or hypervolume (HV) [17] computes the size of the region that is dominated by a set of non-dominated solutions, based on a reference vector that is constructed using the worst objective values of each objective. It is defined as

$$HV(PF) = \cup_{f \in PF} HV(\mathbf{f}) \ with \ HV(\mathbf{f}) = \{\mathbf{f}' \in O : \mathbf{f} \prec \mathbf{f}'\} \tag{5.8}$$

where PF denotes the set of non-dominated sets, O is the objective space and $HV(\mathbf{f})$ is the set of objective vectors dominated by $\mathbf{f}$.

In order to compare one algorithm against another, the HV metric is calculated for each iteration just before a change in the environment occurs. The average over 30 runs is then calculated for each of these iterations.

If it is unknown when a change will occur, the performance metrics can be calculated over all iterations instead of only the iterations just before a change occurrs in the environment.

To determine whether there is a significant difference in the performance of one algorithm compared to another algorithm, statistical tests are used as explained in Section 5.4.3.

5.4.3 Statistical Analysis

To determine whether there is a difference in performance with respect to the performance metrics, a Kruskal-Wallis test was performed for each function. If this test indicates that there is a difference, pairwise Mann-Whitney tests were performed. In all of these tests the average hypervolume value and average spacing value for each of the variations as indicated in Tables 5.2, 5.4, 5.6 and 5.8 were compaired against the average hypervolume value and average spacing value for each of the variations as indicated in Tables 5.3, 5.5, 5.7 and 5.9. The p-values of these tests can be seen in Table 5.1.

Table 5.1. p-values of Statistical Tests

Function	Kruskal-Wallis		Mann-Whitney	
	$\overline{S}$	$\overline{HV}$	$\overline{S}$	$\overline{HV}$
FDA1	0.73	0.6056	–	–
FDA2	0.9384	0.8519	–	–
FDA4	**1.237×10^{-8}**	**9.54×10^{-6}**	**1.101×10^{-8}**	**4.93×10^{-6}**
FDA5	**0.003097**	**0.0001263**	**0.009095**	**7.868×10^{-5}**

5.5 Results

This section discusses the results that were obtained from the experiments, with regards to the performance of the variations of VEPSO and the effect of the way in which knowledge is transferred on VEPSO's performance. Furthermore, the effect of the population size and response to a detected change in the environment is highlighted. The results of the experiments can be seen in Tables 5.2–5.9 and Figs. 5.1–5.4. The overall performance of the knowledge transfer strategies can be seen in Table 5.10. In the tables A indicates that all swarms are re-initialised in response to a change and C indicates that only the swarms that solve the objective functions that have changed are re-initialised. The values that are printed in bold in all tables indicate the best value for the specific metric. In all figures the filled "$\diamond$" symbol indicates solutions found when 20% of the particles, of only the swarms whose objective function changed, are re-initialised and the "$\times$" symbol indicates solutions found when 20% of the particles of all swarms are re-initialised when a change is detected in the environment (refer to Sect. 5.5.3).

5.5.1 Overall Performance

Table 5.10 highlights the overall performance of VEPSO when using either a ring topology or a random topology to exchange knowledge between the swarms. For the 2-objective functions, namely FDA1 and FDA2, there are no real statistical significant differences in VEPSO's performance with regards to the *spacing* and *hypervolume* performance metrics. However, using the random topology for knowledge exchange does lead to a wider spread of solutions for FDA2 as can be seen in Fig. 5.2. For the 3-objective functions, FDA4 and FDA5, there is a significant difference in performance. This is indicated by the p-values in Table 5.1. The p-values that indicate a significant difference in performance are highlighted in bold. From Table 5.10 it can be seen that the random topology lead to a much higher hypervolume value for FDA4, but a higher spacing metric value. For FDA5, the random topology lead to a lower spacing metric value, but a lower hypervolume value. This indicates that more investigation is required to determine the effect on performance for a wider range of functions. Furthermore, randomly selecting

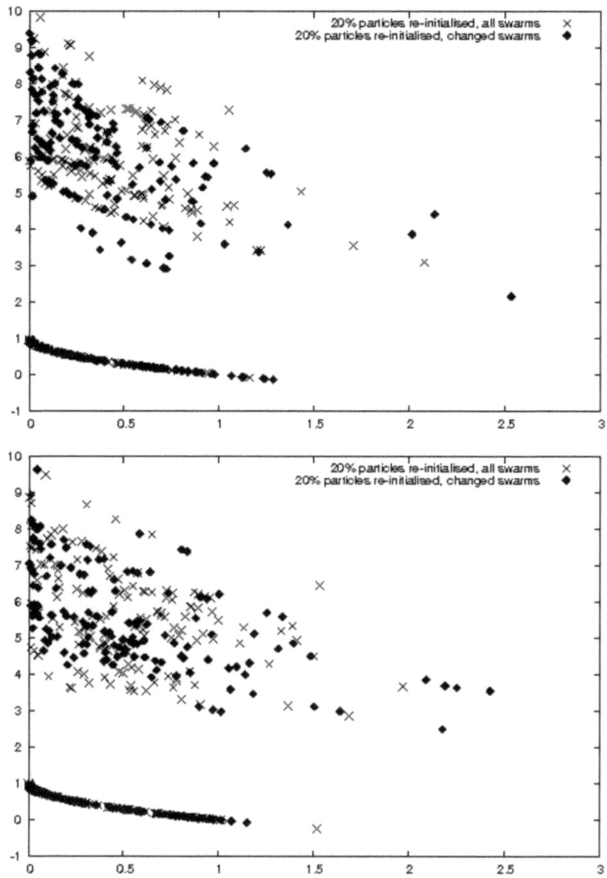

Fig. 5.1. Solutions for Function FDA1 using ring topology on the top and random topology on the bottom

Table 5.2. Spacing and Hypervolume Metric Values for Function FDA1

%R	Ring Topology: #Particles									
	10		20		30		40		Avg	
	$\overline{S}$	$\overline{HV}$	$\overline{S}$	$\overline{HV}$	$\overline{S}$	$\overline{HV}$	$\overline{S}$	$\overline{HV}$	$\overline{S}$	$\overline{HV}$
10A	0.362	6.55	0.35	9.04	0.349	8.58	0.359	10.14	0.355	8.578
10C	0.386	8.27	0.3	8.79	0.345	8.22	0.361	9.72	**0.348**	8.75
20A	0.379	8.71	**0.274**	6.98	0.341	8.32	0.332	8.61	0.332	8.155
20C	0.343	10.24	0.399	10.16	0.349	8.34	0.351	8.34	0.361	9.27
30A	0.355	8.64	0.352	9.23	0.35	**10.92**	0.392	9.08	0.362	9.468
30C	0.381	10.67	0.362	10.12	0.39	10.12	0.362	8.29	0.373	**9.8**
Avg	0.368	8.847	**0.34**	**9.053**	0.354	7.693	0.36	9.03	–	–

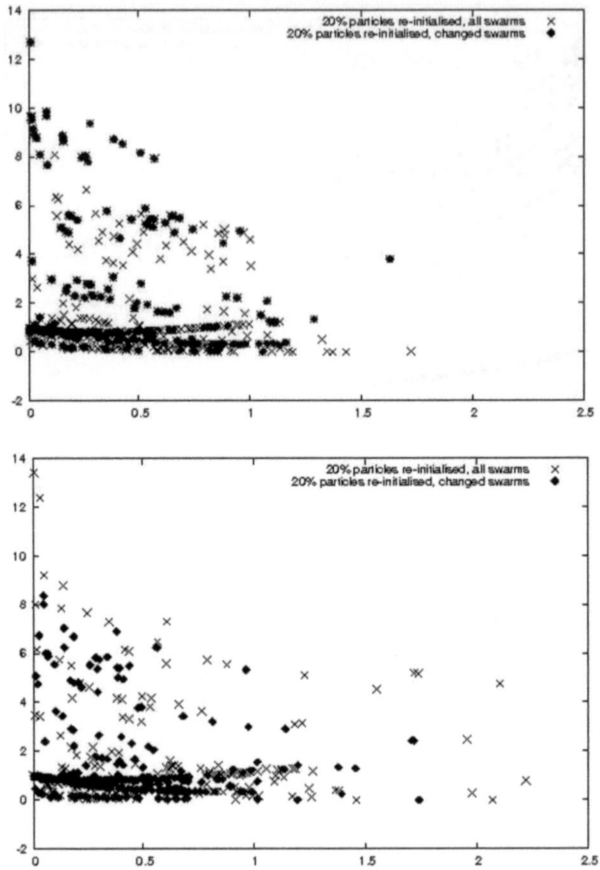

Fig. 5.2. Solutions for Function FDA2 using ring topology on the top and random topology on the bottom

Table 5.3. Spacing and Hypervolume Metric Values for Function FDA1

%R	Random Topology: #Particles								Avg	
	10		20		30		40			
	$\overline{S}$	$\overline{HV}$	$\overline{S}$	$\overline{HV}$	$\overline{S}$	$\overline{HV}$	$\overline{S}$	$\overline{HV}$	$\overline{S}$	$\overline{HV}$
10A	0.37	9.272	0.326	9.531	0.35	7.481	0.342	8.35	**0.347**	8.659
10C	0.353	7.909	0.369	**10.19**	0.352	7.512	0.331	9.165	0.351	8.694
20A	0.341	7.829	0.365	9.944	0.377	8.686	0.362	9.352	0.361	8.953
20C	0.373	9.278	0.353	8.805	0.399	8.676	0.375	10.09	0.375	9.212
30A	0.339	8.453	0.363	8.781	0.384	8.835	0.351	7.768	0.359	8.459
30C	0.444	9.114	0.335	9.393	0.339	9.482	**0.324**	8.971	0.361	**9.24**
Avg	0.37	8.643	0.352	**9.441**	0.367	8.445	**0.348**	8.949	–	–

Table 5.4. Spacing and Hypervolume Metric Values for Function FDA2

%R	Ring Topology: #Particles									
	10		20		30		40		Avg	
	$\overline{S}$	$\overline{HV}$	$\overline{S}$	$\overline{HV}$	$\overline{S}$	$\overline{HV}$	$\overline{S}$	$\overline{HV}$	$\overline{S}$	$\overline{HV}$
10 A	0.264	3.411	0.235	4.293	0.2	4.762	0.307	4.79	0.252	4.314
10 C	0.236	5.155	0.487	9.561	0.238	4.335	0.317	5.503	0.32	4.092
20 A	0.33	5.957	**0.146**	3.034	0.269	4.574	0.223	3.106	**0.161**	4.17
20 C	0.267	3.396	0.236	4.25	0.256	4.574	0.264	3.862	0.256	4.021
30 A	0.175	2.828	0.24	4.905	0.25	3.13	0.262	3.214	0.232	3.519
30 C	0.207	3.958	0.243	4.399	0.248	**15.995**	0.253	5.74	0.238	**5.015**
Avg	0.247	4.118	0.265	5.074	**0.244**	**6.228**	0.271	4.369	–	–

Table 5.5. Spacing and Hypervolume Metric Values for Function FDA2

%R	Random Topology: #Particles									
	10		20		30		40		Avg	
	$\overline{S}$	$\overline{HV}$	$\overline{S}$	$\overline{HV}$	$\overline{S}$	$\overline{HV}$	$\overline{S}$	$\overline{HV}$	$\overline{S}$	$\overline{HV}$
10 A	0.251	3.963	0.253	4.513	0.219	2.357	0.284	**10.77**	0.252	5.401
10 C	0.365	8.9	0.476	5.804	0.259	4.444	0.317	6.391	0.354	**6.385**
20 A	0.3	4.426	0.172	3.315	0.338	6.904	0.548	7.208	0.226	5.463
20 C	0.247	4.028	0.193	2.489	0.263	4.71	0.215	4.881	**0.153**	4.027
30 A	0.216	4.625	**0.145**	2.745	0.244	5.105	0.205	3.313	0.203	3.947
30 C	0.214	3.954	0.241	3.566	0.195	2.735	0.482	5.583	0.283	3.957
Avg	0.266	4.983	**0.247**	3.739	0.253	4.376	0.341	**6.358**	–	–

Table 5.6. Spacing and Hypervolume Metric Values for Function FDA4

%R	Ring Topology: #Particles									
	10		20		30		40		Avg	
	$\overline{S}$	$\overline{HV}$	$\overline{S}$	$\overline{HV}$	$\overline{S}$	$\overline{HV}$	$\overline{S}$	$\overline{HV}$	$\overline{S}$	$\overline{HV}$
10 A	0.19	1.08	0.19	0.39	0.19	0.24	0.03	1.19	0.15	0.72
10 C	0.03	6.04	0.02	1.3	0.09	1230.2	0.14	**5122.9**	0.07	1590.1
20 A	0.03	9.25	0.03	0.33	0.03	1.47	0.03	0.43	**0.03**	2.87
20 C	0.04	2.60	0.03	3.22	0.03	0.71	0.03	5.62	0.03	3.04
30 A	0.03	2.64	0.02	0.37	0.02	0.49	0.08	170.27	0.04	43.45
30 C	**0.02**	0.09	0.03	9.63	0.03	1.22	0.13	4107.2	0.05	**1029.5**
Avg	0.06	3.62	**0.03**	2.54	0.06	205.72	0.07	**1567.9**	–	–

which swarm's knowledge to use (random topology) leads to a wider spread of solutions for FDA4, as can be seen in Fig. 5.3.

5.5.2 Population Size

From Tables 5.2–5.9 it is interesting to note that for the ring topology the best hypervolume values for functions FDA1, FDA4 and FDA5 were obtained

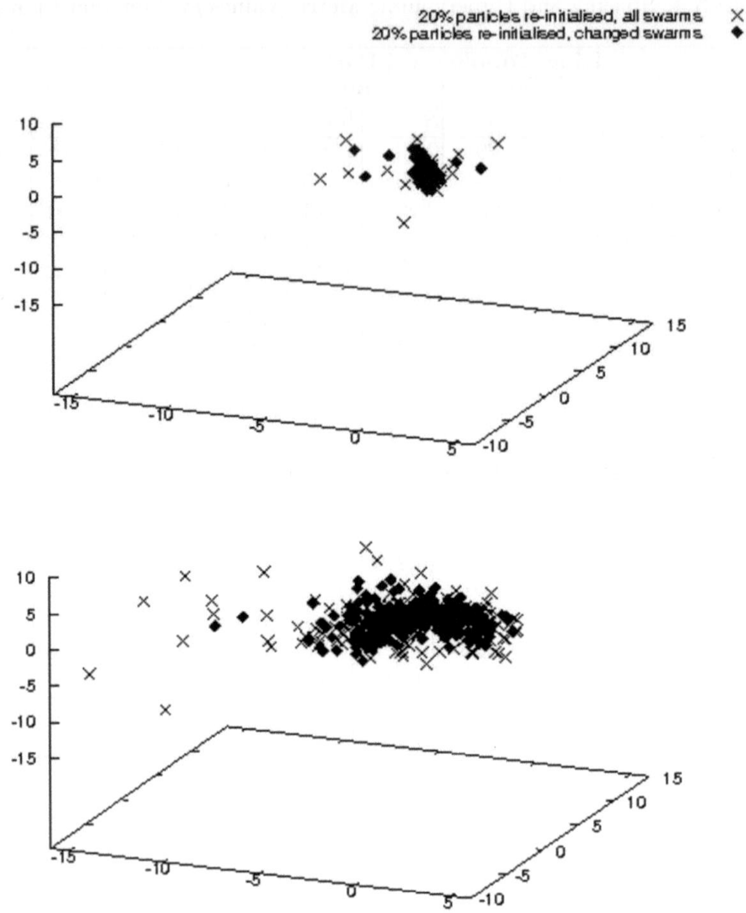

Fig. 5.3. Solutions for Function FDA4 using ring topology on the top and random topology on the bottom

Table 5.7. Spacing and Hypervolume Metric Values for Function FDA4

%R	Random Topology: #Particles									
	10		20		30		40		Avg	
	$\overline{S}$	$\overline{HV}$	$\overline{S}$	$\overline{HV}$	$\overline{S}$	$\overline{HV}$	$\overline{S}$	$\overline{HV}$	$\overline{S}$	$\overline{HV}$
10 A	0.21	249.56	0.24	274.31	0.37	4580.01	0.27	1241.37	0.27	1586.31
10 C	0.46	**60434.4**	0.20	114.76	0.28	600.62	0.19	59.5	0.28	**15302.3**
20 A	0.21	97.98	0.20	2054.1	0.31	100.02	0.3	3080.7	**0.25**	1333.2
20 C	0.19	97.98	0.32	2054.1	0.19	100.02	0.39	3080.70	0.27	1333.2
30 A	0.31	1121.8	0.17	35.95	0.27	277.75	0.31	4503.9	0.26	1484.9
30 C	0.33	1093.8	0.19	42.02	0.4	6379.8	**0.16**	25.31	0.27	1885.2
Avg	0.28	10515.9	**0.22**	762.52	0.3	**12038.2**	0.27	1998.6	–	–

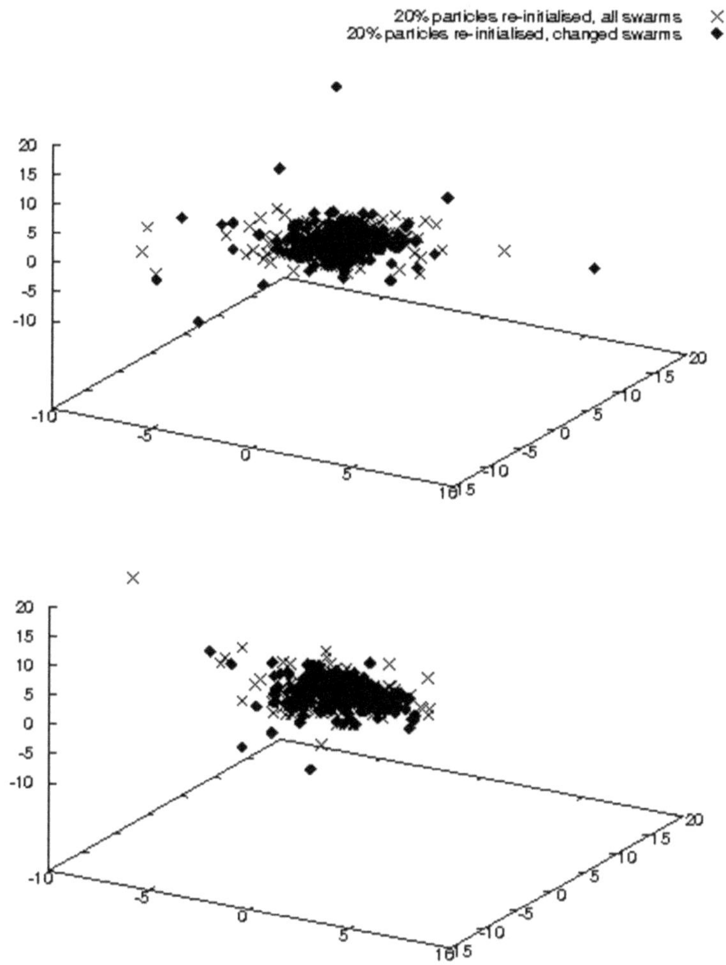

Fig. 5.4. Solutions for Function FDA5 using ring topology on the top and random topology on the bottom

when each swarm had 20 particles. For the random topology the highest hypervolume values for functions FDA2, FDA4 and FDA5 were obtained using 40 particles per swarm. When knowledge was exchanged using the ring topology, swarms that consisted of 30 particles lead to the lowest spacing values for functions FDA2 and FDA5. However, for FDA1 and FDA4, the best values for the spacing metric were obtained with 20 and 10 particles respectively. When a random topology was used to exchange knowledge between the swarms, the best spacing values were obtained for functions FDA1, FDA2 and FDA4 using 20 particles per swarm. For FDA5 30 particles per swarm lead to the best spacing values.

Table 5.8. Spacing and Hypervolume Metric Values for Function FDA5

%R	Ring Topology: #Particles									
	10		20		30		40		Avg	
	$\overline{S}$	$\overline{HV}$	$\overline{S}$	$\overline{HV}$	$\overline{S}$	$\overline{HV}$	$\overline{S}$	$\overline{HV}$	$\overline{S}$	$\overline{HV}$
10 A	**0.12**	67.0	0.4	77.47	0.31	35.99	0.32	41.54	0.31	55.49
10 C	0.61	52.94	0.29	201.35	0.78	1672.76	0.68	3334.49	0.59	1315.39
20 A	0.55	6207.04	0.53	**233672.3**	0.29	20336.39	0.63	41.75	0.5	**65064.4**
20 C	0.54	30274.45	0.48	285.33	0.41	4882.7	0.37	422.62	0.45	8966.27
30 A	0.47	257.3	0.49	12231.54	0.54	379.1	0.79	3844.01	0.57	4177.99
30 C	0.63	2185.09	0.37	249.35	0.4	478.29	0.24	452.92	**0.27**	841.41
Avg	0.5	6507.3	0.43	**41119.55**	**0.32**	4630.86	0.50	1356.23	–	–

Table 5.9. Spacing and Hypervolume Metric Values for Function FDA5

%R	Random Topology: #Particles									
	10		20		30		40		Avg	
	$\overline{S}$	$\overline{HV}$	$\overline{S}$	$\overline{HV}$	$\overline{S}$	$\overline{HV}$	$\overline{S}$	$\overline{HV}$	$\overline{S}$	$\overline{HV}$
10 A	0.41	53.53	0.38	48.13	0.25	23.47	0.32	22.14	0.34	36.82
10 C	0.37	29.51	0.3	415.95	0.33	11.97	0.94	**185921.7**	0.485	**46594.8**
20 A	0.41	14.71	0.62	205.28	**0.16**	26.92	0.27	21.95	0.37	67.21
20 C	0.40	1092.41	0.32	110.61	0.45	275.92	0.45	169.91	0.51	412.21
30 A	0.28	13.11	0.19	28.81	0.16	6.83	0.32	149.66	**0.24**	49.61
30 C	0.26	260.53	0.29	113.73	0.33	24.25	0.37	14.95	0.31	103.37
Avg	0.35	243.97	0.35	153.75	**0.28**	61.56	0.45	**31050.1**	–	–

Table 5.10. Overall Result Comparison

Function	Ring Topology		Random Topology	
	$\overline{S}$	$\overline{HV}$	$\overline{S}$	$\overline{HV}$
FDA1	**0.355**	**9.003**	0.359	8.869
FDA2	**0.256**	**4.947**	0.277	4.864
FDA4	**0.063**	444.953	0.268	**3820.845**
FDA5	0.47	**13403.485**	**0.357**	7877.332

5.5.3 Response Strategies

When using a ring topology for knowledge transfer, the best hypervolume values for FDA2, FDA4 and FDA5 were obtained when 30% of the particles were re-initialised of the swarm that is optimising the objective function that has changed. However, when a random topology is used to transfer knowledge between the swarms, the best hypervolume values were obtained when 10% of the particles were re-initialised of the swarm that is optimising the objective function that has changed (refer to Tables 5.2–5.9). The best spacing values were obtained for both functions FDA2 and FDA4 re-initialising 20% of the particles of all swarms. For FDA1 and FDA5 the best spacing values were obtained by re-initialising 10% and 30% of the particles of only the swarm whose objective function changed respectively.

5.6 Summary

DMOO problems occur in a vast range of situations in modern life. The objectives are in conflict with one another and a change in the environment influences the solutions of the optimisation problem. This chapter adapted the VEPSO approach to solve DMOO problems. Two approaches to transfer knowledge between the swarms were presented, namely a ring or random topology. Furthermore, the performance of VEPSO was highlighted using these two knowledge exchange approaches when the number of particles in the swarms, and the response to a change is varied. Results indicated that there is not a statistical significant difference in the performance of VEPSO when these two knowledge exchange approaches were used for problems with only 2 objectives. However, when the problem has 3 objectives the random knowledge transfer topology leads to an improvement in either the hypervolume value or the spacing value. Further investigations will be done to determine the effect with a wider range of functions, with a various number of objectives.

Current research focuses on improving the performance of the VEPSO approach for DMOOPs. However, the best version of VEPSO will be compared against other approaches in the future.

References

1. Greeff, M., Engelbrecht, A.P.: Solving Dynamic Multi-objective Problems with Vector Evaluated Particle Swarm Optimisation. In: Proc. of IEEE World Congress on Computational Intelligence (WCCI): IEEE Congress on Evolutionary Computation (CEC), June 2008, pp. 2917–2924 (2008)
2. Kennedy, J., Eberhart, R.C.: Particle Swarm Optimization. In: Proc. of IEEE International Conference on Neural Networks, vol. IV, pp. 1942–1948 (1995)
3. Li, X., Dam, K.H.: Comparing Particle Swarms for Tracking Extrema in Dynamic Environments. In: Proc. of Congress on Evolutionary Computation (CEC 2003), December 2003, vol. 3, pp. 1772–1779 (2003)
4. Li, X., Branke, J., Blackwell, T.: Particle Swarm with Speciation and Adaptation in a Dynamic Environment. In: Proc. of 8th Conference on Genetic and Evolutionary Computation (GECCO 2006), pp. 51–58 (2006)
5. Blackwell, T., Branke, J.: Multi-swarm Optimization in Dynamic Environments. In: Raidl, G.R., Cagnoni, S., Branke, J., Corne, D.W., Drechsler, R., Jin, Y., Johnson, C.G., Machado, P., Marchiori, E., Rothlauf, F., Smith, G.D., Squillero, G. (eds.) EvoWorkshops 2004. LNCS, vol. 3005, pp. 489–500. Springer, Heidelberg (2004)
6. Cui, X., et al.: Tracking Non-Stationary Optimal Solution by Particle Swarm Optimizer. In: Proc. of the 6th International Conference on Software Engineering, Artificial Intelligence, Networking and Parallel/Distributed Computing and First ACIS International Workshop on Self-Assembling Wireless Networks (SNPD/SAWN 2005), May 2005, pp. 133–138 (2005)

7. Carlisle, A., Dozier, G.: Adapting Particle Swarm Optimization to Dynamic Environments. In: Proc. of International Conference on Artificial Intelligence (ICAI 2000), pp. 429–434 (2000)
8. Hu, X., Eberhart, R.C.: Adaptive Particle Swarm Optimization: Detection and Response to Dynamic Systems. In: Proc. of IEEE Congress on Evolutionary Computation (CEC 2002) (May 2002)
9. Carlisle, A., Dozier, G.: Tracking changing extrema with adaptive particle swarm optimizer. In: Proc. of 5th Biannual World Automation Congress, vol. 13, pp. 265–270 (2002)
10. Reyes-Sierra, M., Coello Coello, C.A.: Multi-Objective Particle Swarm Optimizers: A Survey of the State-of-the-Art. International Journal of Computational Intelligence Research 2(3), 287–308 (2006)
11. Guan, S.-U., Chen, Q., Mo, W.: Evolving Dynamic Multi-Objective Optimization Problems with Objective Replacement. Artificial Intelligence Review 23(3), 267–293 (2005)
12. Mehnen, J., Wagner, T., Rudolph, G.: Evolutionary Optimization of Dynamic Muli-Objective Test Functions. In: Proc. of 2nd Italian Workshop on Evolutionary Computation and 3rd Italian Workshop on Artificial Life (2006)
13. Hatzakis, I., Wallace, D.: Dynamic Multi-Objective Optimization with Evolutionary Algorithms: A Forward Looking Approach. In: Proc. of 8th Annual Conference on Genetic and Evolutionary Computation (GECCO 2006), vol. 2, pp. 1201–1208 (2006)
14. Li, X., Branke, J., Kirley, M.: On Performance Metrics and Particle Swarm Methods for Dynamic Multiobjective Optimization Problems. In: Proc. of Congress of Evolutionary Computation (CEC 2007), pp. 1635–1643 (2007)
15. Cámara, M., Ortega, J., Toro, J.: Parallel Processing for Multi-objective Optimization in Dynamic Environments. In: Proc. of IEEE International Parallel and Distributed Processing Symposium, p. 243 (2007)
16. Jin, Y., Sendhoff, B.: Constructing Dynamic Optimization Test Problems using the Multi-objective Optimization Concept. In: Raidl, G.R., Cagnoni, S., Branke, J., Corne, D.W., Drechsler, R., Jin, Y., Johnson, C.G., Machado, P., Marchiori, E., Rothlauf, F., Smith, G.D., Squillero, G. (eds.) EvoWorkshops 2004. LNCS, vol. 3005, pp. 525–536. Springer, Heidelberg (2004)
17. Zitzler, E., Deb, K., Thiele, L.: Comparison of Multiobjective Evolutionary Algorithms: Empirical Results. Evolutionary Computation 8(2), 173–195 (2000)
18. Deb, K., et al.: Scalable multi-objective optimization test problems. In: Proc. of Congress on Evolutionary Computation (CEC 2002), vol. 1, pp. 825–830 (2002)
19. Farina, M., Deb, K., Amato, P.: Dynamic multiobjective optimization problems: test cases, approximations, and applications. IEEE Transactions on Evolutionary Computation 8(5), 425–442 (2004)
20. Zheng, B.: A New Dynamic Multi-Objective Optimization Evolutionary Algorithm. In: Proc. of third International Conference on Natural Computation (ICNC 2007), vol. V, pp. 565–570 (2007)
21. Schaffer, J.D.: Multiple Objective Optimization with Vector Evaluated Genetic Algorithms. In: Proc. of the 1st Intenational Conference on Genetic Algorithms, pp. 93–100 (1985)
22. Fieldsend, J.E., Everson, R.M.: Multi-objective Optimisation in the Presence of Uncertainty. In: Proc. of IEEE Congress on Evolutionary Computation, pp. 476–483 (2005)

23. Goh, C.K., Tan, K.C.: An Investigation on Noisy Environments in Evolutionary Multiobjective Optimization. IEEE Transactions on Evolutionary Computation 11(3), 354–381 (2007)
24. van Veldhuizen, D.A., Lamont, G.B.: On Measuring Multiobjective Evolutionary Algorithm Performance. In: Proc. of Congress on Evolutionary Computation (CEC 2000), July 2000, pp. 204–211 (2000)
25. Zheng, S.-Y.: et al. A Dynamic Multi-Objective Evolutionary Algorithm Based on an Orthogonal Design. In: Proc. of IEEE Congress on Evolutionary Computation (CEC 2006), July 2006, pp. 573–580 (2006)
26. Parsopoulos, K.E., Tasoulis, D.K., Vrahatis, M.N.: Multiobjective Optimization using Parallel Vector Evaluated Particle Swarm Optimization. In: Proc. of IASTED International Conference on Artificial Intelligence and Applications (2004)
27. Parsopoulos, K.E., Vrahatis, M.N.: Recent Approaches to Global Optimization Problems through Particle Swarm Optimization. Natural Computing 1(2-3), 235–306 (2002)
28. Bartz-Beielstein, T., et al.: Particle Swarm Optimizers for Pareto Optimization with Enhanced Archiving Techniques. In: Proc. Congress on Evolutionary Computation (CEC), pp. 1780–1787 (2003)
29. Bergh, F.v.d.: An analysis of particle swarm optimizers. PhD thesis, Department of Computer Science, University of Pretoria (2002)

6

Meta-PSO for Multi-Objective EM Problems

Marco Mussetta[1,3], Paola Pirinoli[1], Stefano Selleri[2], and Riccardo E. Zich[3]

[1] Politecnico di Torino, Department of Electronics,
Corso Duca degli Abruzzi 24, I-10129 Torino, Italy
marco.mussetta@polito.it, paola.pirinoli@polito.it
[2] University of Florence, Department of Electronics and Telecommunications,
Via C. Lombroso 6/17, I-50134 Florence, Italy
stefano.selleri@unifi.it
[3] Politecnico di Milano, Department of Energy,
Piazza Leonardo da Vinci 32, I-20133 Milano, Italy
riccardo.zich@polimi.it

Recently, the Particle Swarm Optimization (PSO) method has been successfully applied to many different electromagnetic optimization problems. Due to the complex equations, usually calling for a numerical solution, the associated cost function is in general very computationally expensive. A fast convergence of the optimization algorithm is hence of paramount importance to attain results in an acceptable time.

In this chapter few variations over the standard PSO algorithm, referred to as Meta-PSO, aimed to enhance the global search capability, and, therefore, to improve the algorithm convergence, are analyzed.

The Meta-PSO class of methods can be furthermore subdivided into the Undifferentiated and a Differentiated subclasses, whether the law updating particle velocity is the same for all particles or not, respectively.

In recently published open literature the results of the application of the Meta-PSO to the optimization of single-objective problems have been shown. Here we will prove their enhanced properties with respect to standard PSO also for the optimization of multi-objective problems, trough their test in multi-objective benchmarks and multi-objective optimization of an antenna array.

6.1 Introduction

Dealing with real life optimization problems, as practical applied engineering problems and more specifically electromagnetic ones, it is rather common to have to handle problems that intrinsically present more than one objective function.

In such problems, the objectives to be optimized are typically in conflict one with respect to each other, and this means that trying to define what

N. Nedjah et al. (Eds.): Multi-Objective Swarm Intelligent Systems, SCI 261, pp. 125–150.
springerlink.com

is the single optimal solution for this class of problems is pointless. On the contrary, the challenge is to find good "trade-off" solutions that represent the best possible compromises among all the objectives, extending the concept of best solution to multi-objective problems.

In view of this, multi-objective optimization may be defined as a strategy to address multiple design constrains in practical engineering problems. Several approaches have been proposed to address this kind of optimization problems [1, 2]. Among them, one of the most commonly adopted is the so called weighted sum method (WSM); it consists in rephrasing the multi-objective optimization problem in a suitably defined equivalent single-objective one, whose fitness function is the linear combination of several separate fitness functions $f_k(\mathbf{x})$, each of which takes into account only one of the design constrains. This approach is equivalent to projecting the hyper-search-space on a suitable plane, and the resulting overall fitness function has the form

$$f(\mathbf{x}) = \sum_{k=1}^{K} w_k f_k(\mathbf{x}) \tag{6.1}$$

where w_k is the weight that the kth constrains has with respect to the others. Usually the weighting coefficients are chosen in order to satisfy the normalization relationship

$$\sum_{k=1}^{K} w_k = 1 \tag{6.2}$$

and by changing their values it is possible to control the relative importance of each design constrain, obtaining consequently different overall fitness functions and sets of solutions.

The key issue of the WSM approach is therefore to find out the best trade-off between all the weighting coefficients and this requires a good knowledge of the relative importance of each of them with respect to the others. For this reason WSM often requires an extensive tuning of the weighting coefficients w_k, particularly for problems where objectives are unrelated. Another issue related to this approach is the inability to find the Pareto-optimal solutions in non-convex regions (as will be explained in the following), although a solution of this approach is always Pareto-optimal.

Another approach traditionally adopted to handle multi-objective problems is the so called ε-constrained method (εCM) [2], according to which all the objectives are constrained but one, so that, among the several $f_k(\mathbf{x})$ to be optimized, the task is reduced to the minimization of $f_\mu(\mathbf{x})$, subject to $f_k(\mathbf{x}) \leq \epsilon_k$, when $k \neq \mu$. An example of this approach has already been considered by the authors in the cost function definition of [3]. The difficulties with εCM are the need to know relevant ε vectors and the non-uniformity in Pareto-optimal solutions, although any Pareto-optimal solution can be found with this approach.

The two approaches mentioned above are however merely extensions and adaptations of the single-objective oriented strategy to multi-objective problems. Therefore, in order that the optimization process works well it is necessary to have an in-deep knowledge of the problem to be optimized, so that the weighting coefficients are suitably chosen, and to run the single-objective optimizer a large number of times, to explore the different solutions: even then, good distribution of the results is not guaranteed.

Among the different techniques originally developed for single-objective optimization that can be successfully adopted also in multi-objective problems there is the Particle Swarm Optimization (PSO) algorithm [4, 5]. The PSO gained immediate popularity, especially because of its simplicity in implementation, its robustness and its “optimization capability” for both single-objective and multi-objective problems. PSO proved to be in all considered cases at least comparable, and often superior, to GAs [6]-[10]. Besides the easiness of the implementation, the PSO presents the advantage of being well suited for optimization problems with both discrete (binary and ternary) and continuous parameters, and for parallel computing implementation [11]-[17].

In most applications of the PSO scheme to the optimization of complex engineering problems, the single-objective approach is generally used, and the overall fitness function is generated adopting one of the two method mentioned above. This is also what it is done in optimization of electromagnetic problems (see e.g. [8]-[18]), even if in most cases they present several competing targets. In these case the use of a “true” multi-objective optimization scheme would be therefore convenient.

A first attempt to extend the use of PSO to multi-objective optimization is that proposed in [19] and named vector-evaluated PSO (VEPSO), which utilizes a separate swarm for each design goal and each swarm has its own global best. After each iteration, the global bests of all swarms are exchanged. However, if there are more than two swarms operating in the solution space, it is not obvious how the global bests should be exchanged.

Another implementations of a multi-objective version of the PSO, here called *directed multi-objective PSO*, is that based on the concept of Pareto dominance [20], usually adopted for the definition of the multi-objective optimization algorithm: given two arbitrary vectors $\mathbf{x_1}$ and $\mathbf{x_2}$, we say that $\mathbf{x_1}$ dominates $\mathbf{x_2}$ if and only if their fitness vectors satisfy $f_i(\mathbf{x_1}) \leq f_i(\mathbf{x_2}), \forall i$ and $f_i(\mathbf{x_1}) < f_i(\mathbf{x_2})$ for at least one i. This means that $\mathbf{x_1}$ dominates $\mathbf{x_2}$ if $\mathbf{x_1}$ is not worse than $\mathbf{x_2}$ in all objectives $f_i(\mathbf{x})$ and $\mathbf{x_1}$ is strictly better than $\mathbf{x_2}$ in at least one objective.

A vector $\mathbf{x_1}$ is therefore *non-dominated* if there does not exist another $\mathbf{x_2}$ that dominates it. In this case, $\mathbf{x_1}$ is a Pareto-optimal solution and the set of all non dominated solutions is called the Pareto front. Figure 6.1 shows a particular case of the dominance relation and of the Pareto front in presence of two objective functions.

Both the single-objective and the multi-objective versions of the PSO, as any other optimization technique, require the evaluation of the cost function

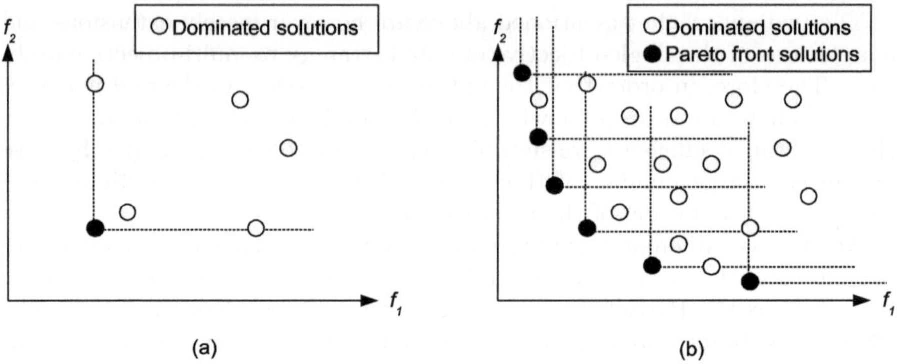

Fig. 6.1. Dominance relation (a) and Pareto front for a set of solutions (b) in a multi-objective space.

thousands of times. In electromagnetic problems, the single or multi-objective cost function is generally very computationally expensive and therefore the cost of the optimization process may become unaffordable. For this reason, in the last years the authors have introduced new versions of the PSO algorithm, with enhanced properties: in [3, 10, 13, 18] some variations of the standard PSO, named Meta-PSO, exploiting multiple interacting swarms and aimed to improve the PSO performances with a negligible overhead in algorithm complexity and computational cost, have been introduced.

In literature, the use of more than one swarm is very sporadic, and, to the authors best knowledge, was completely absent in conjunction with electromagnetic problems up to the previously cited papers and up to a very recent paper [21] where multiple swarm are defined on subdomains, let to evolve and then joined in a single large swarm.

Previous approaches to multiple swarm optimization outside the electromagnetic area can nevertheless be found in [22], where a division of the population in cluster has been proposed. Contrarily to what is proposed here, in [22] the particles belonging to a cluster are chosen according to a "minimum distance" criterion, and the equations that manage the evolution of the single particle are modified, substituting its personal and/or the global bests with those of the center of the cluster the particle belong to. On the other hand in our algorithms swarms composition do not vary, and a particle always belong to the same swarm, allowing for intermingling swarms which nevertheless remains distinct. In this way the complexity of the algorithm remains almost unchanged, whereas in [22] it increases since the particles division into clusters must be performed and managed continuously.

As another multiple-swarm example, in [23] two separate PSOs are used to optimize a two-objective problem: each PSO focus on one aspect of the problem and they interact through the cost function. Finally, in [24] the Cooperative Particle Swarm (CPSO) methods are introduced. In its basic version, this technique splits the domain in subspaces, each searched by a swarm. This approach requires additional computations to reconstruct the point where the cost function has to be evaluated and a more complex way, with respect to conventional PSO, to handle and store personal and global bests. The most performing of the CPSO techniques, the CPSO-H_k, relies on the co-evolution of a CPSO and a PSO with exchange of information between the two, leading to an even more complex algorithm [24].

In opposite to the CPSO, the multiple swarm approaches introduced in [3] use several swarms to span the whole domain in order to have better and faster exploitation of it with respect to a single swarm. The three techniques presented in [3] differ in the rule that manages particles, exhibiting just one or two additional terms in the velocity update function with respect to a standard PSO. The added complexity is hence negligible and of course the intrinsic parallel nature of PSO is maintained if not enhanced: Each swarm can be assigned to a separate process with very little inter-process communication. These three methods will be identified in the following as Undifferentiated Meta-PSOs (U-MetaPSO), the word Undifferentiated meaning that all the particles of all swarms obeys to the same rules.

The simplest of all the U-MetaPSOs was further modified in [18], and two other new schemes have been obtained, named Differentiated Meta-PSOs (D-MetaPSO), in which the particles within a swarm are not managed with the same rule but, indeed, all particles but one obeys to the simplest PSO rules whereas a single particle in the swarm, named *leader* is managed with different rules taking into account inter-swarm relations.

Despite of their simplicity, both the Undifferentiated and the Differentiated Meta-PSO work better than the standard PSO, as appears from results in [3],[18] and as it will be shown here, where they will be applied to the optimization of several multi-objective test functions and to the design of a linear array. The effectiveness of the Undifferentiated and Differentiated Meta-PSO algorithm can also find a confirmation in the analysis on the different possible social interaction models reported in [25], in which it is hypothesized that tightly connected particle swarms, as in the standard PSO scheme, may not be so good in finding the problem optimum, since their tendency to behave as a swarm, which is indeed its main characteristic, lead to the tendency for the particles to stay close together and hence to a non-negligible possibility of being entrapped in local sub-optima, while this risk is lower in case of moderately connected societies, as the here proposed schemes are. This observation is confirmed by the analysis of the reliability of the solution presented in [3].

In all the applications presented in [3, 10, 13, 18] the problem to be optimized has been reduced to a single-objective problem. Here, the Meta-PSO algorithm will be extended to the optimization of Multi-Objective (MO) problems. Their application to both test functions and real-life problems will show that they outperform the standard Multi-Objective PSO, since, thank to their better capability in exploring the domain of definition of the optimization problem, a high number of agents reach the Pareto front.

This chapter is organized as it follows. In Section 6.2 the basic PSO algorithm is briefly sketched for notation consistency and to point out some features - like boundary conditions - which will be useful after on; then in Section 6.3 the three U-MetaPSO are presented and in Section 6.4 the two D-MetaPSO are given. Section 6.5 presents some previously obtained results over single-objective cases to tune the algorithms. In fact, a weak point of most stochastic techniques is that their performances (capability and speed of convergence) strongly depend on the values of the different parametric constants appearing in the algorithm. A fine tuning of the equations controlling the evolution of the particles in the solution space is hence mandatory. In Section 6.6 the extension of the Meta-PSO to multi-objective problems is presented through their application to testing functions that permits to compare the performances of the Meta-PSOs among them and with the standard PSO. In Section 6.7 the results of the application of the multi-objective Meta-PSOs to a real life problem, i.e. the optimization of a linear beam scanning array working on two different frequency bands, are presented, while the closing remarks reported in Section 6.8 conclude the chapter.

6.2 Basic PSO Algorithm

The standard PSO algorithm is an iterative procedure in which a set of particles, or agents, moves in the parameter space, which is a subset $\mathscr{D}$ of a M-dimensional space on which the cost function $\mathscr{F}$ is defined. A full treatment of the method can be found elsewhere in this book and on many papers, as, for example, [26] but for sake of clarity and uniformity of notation it will be briefly summarized in the following too.

Let's assume to have a set of $i = 1, \ldots, N_p$ particles, or agents, each characterized by its position $\mathbf{X}_i$ in the aforementioned M-dimensional space domain $\mathscr{D}$ and by its velocity $\mathbf{V}_i$ with which it moves.

At the beginning of the simulation the set of positions and velocities have completely random values $\mathbf{X}_i^{(0)}$ and $\mathbf{V}_i^{(0)}$. Then, at each iteration, they are updated iteratively according to the rules:

$$\mathbf{V}_i^{(k+1)} = \omega^{(k)}\mathbf{V}_i^{(k)} + \phi\eta_1(\mathbf{P}_i - \mathbf{X}_i^{(k)}) + \phi\eta_2(\mathbf{G} - \mathbf{X}_i^{(k)}) \tag{6.3}$$

$$\mathbf{X}_i^{(k+1)} = \mathbf{X}_i^{(k)} + \mathbf{V}_i^{(k+1)} \tag{6.4}$$

being $\mathbf{P}_i$ the best position ever attained by particle i itself and $\mathbf{G}$ the best position ever attained by the particle swarm; $\omega^{(k)} = \omega_0 e^{-\alpha k} + \omega_1$ is a inertial factor slowing down particles (Figure 6.2).

In this paradigm (6.3) represent a velocity variation, hence an acceleration. If mass is considered equal to one the last two terms in this equation are effectively forces, both of which attracts the particle towards $\mathbf{P}_i$ and $\mathbf{G}$. In this sense the particles exhibit a personal ($\mathbf{P}_i$) and social ($\mathbf{G}$) knowledge. The coefficients

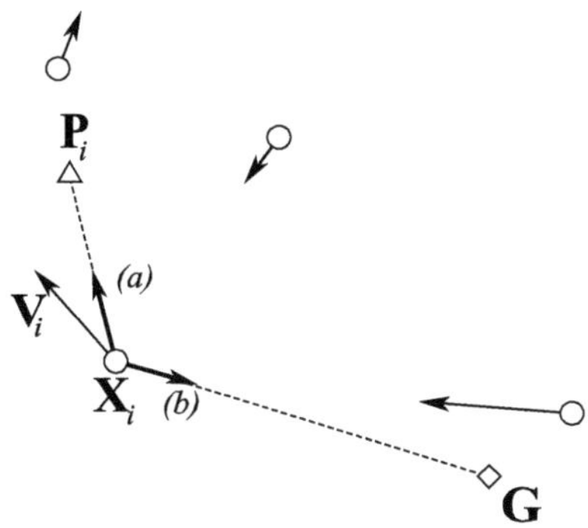

Fig. 6.2. PSO basic layout. Forces over a generic particle i are: (a) pull toward personal best $\mathbf{P}_i$; (b) pull toward global best $\mathbf{G}$.

η_1 and η_2 are positive parameters tuning these pulls towards the personal and global best positions while ϕ is a random number of uniform distribution in the $[0, 1]$ range. Note that, if ϕ appears more that once in a given formula, as in (6.3) it is assumed to have different random value each time.

The inertial factor can be either constant or iteration-dependent. This second possibility was proposed in [27] and has several advantages: starting the optimization process with a high value for $\omega^{(k)}$ and reducing it as k increases encourages the particles to initially explore the whole space domain, in search of the global minimum, and then allow them to better investigate the region in which this minimum is supposed to be located. Inertia is indeed a key issue in PSO based optimization and adaptively setting this inertia via Fuzzy systems has been recently proposed [28].

An important aspect connected with the efficiency of the PSO is the way in which the particles moving towards the border of the solution space are handled. In [8] three different solutions have been proposed: the first one consists in setting to zero the velocity of the particles arriving at the domain boundary, the second one models the boundaries as perfect reflecting surfaces, so that the particles impinging on them are reflected back in the solution space as for a billiard ball hitting the side; finally the third one allows the particles to fly out from the solution space, without evaluating the cost function any more, until the particle eventually gets back in the domain. A further classification of boundary condition is provided in [29], where a fourth condition, damping, is added, consisting in a reflection condition on which a random multiplicative factor smaller than 1 is applied to the new particle velocity. In [30], on the other hand, a periodic condition is introduced, by letting the particle which exits from the domain to maintain its velocity but to be transported on the other side of the domain, hence enhancing domain exploitation. Further choices, like moving back escaping particles to their personal best to allow a finer search around that position were also proposed [31].

The results reported in this Chapter have been obtained adopting the reflecting boundaries, since trials seems to suggest that this choice guarantees the faster convergence.

6.3 Undifferentiated Meta-PSO

Some variations over the standard PSO algorithm briefly described in the previous section will be presented here. As stated in the Introduction all the variations relies on the concept of multiple, mutually-interacting swarms to enhance the algorithm capabilities in the global search in the parameter space.

They are grouped into two classes, Undifferentiated and Differentiated, according to the fact that the velocity update rule is the same for all agents or not, respectively. This section will deal with the Undifferentiated class.

6.3.1 Meta-PSO

Meta-PSO (MPSO) is the most straightforward of the methods here presented and simply consists in using more than a single swarm and by letting information exchange between them. Particles are now characterized by two indices: an index $j = 1, \ldots, N_s$ defining the swarm they belong to and an index $i = 1, \ldots, N_{p_j}$ within the swarm. For sake of simplicity in the following all swarms will be considered as having the same number of particles $N_p = N_{p_j} \forall j = 1, \ldots, N_s$, but this is not required. The MPSO velocity update rule is:

$$\mathbf{V}_{j,i}^{(k+1)} = \omega^{(k)}\mathbf{V}_{j,i}^{(k)} + \phi\eta_1(\mathbf{P}_{j,i} - \mathbf{X}_{j,i}^{(k)}) + \phi\eta_3(\mathbf{S}_j - \mathbf{X}_{j,i}^{(k)}) + \phi\eta_2(\mathbf{G} - \mathbf{X}_{j,i}^{(k)}) \quad (6.5)$$

where $\mathbf{P}_{j,i}$ is the particle personal best position, $\mathbf{S}_j$ is the global best position of swarm j (swarm or social knowledge) and $\mathbf{G}$ is the global best position of all swarms (inter-swarm or racial knowledge), while the other symbols have the same meaning than in (6.3).

The essence of this modification is that of inserting a new global best, which is the best of all the swarm's global best. If the swarm global best can be considered a social knowledge, then this new best position can be assumed to be a *racial knowledge* in the sense that is a knowledge shared by all members of the same kind (or *race*) of particles, forming separate communities (or *swarms*) (Figure 6.3). Each swarm retains its own 'global' best which, to prevent misinterpretations, is here called *swarm* best.

The effectiveness of this approach can be understood by resorting to the concept that loosely connected societies might be better than tightly connected ones in finding a global optimum [25] and this approach, exploiting multiple swarms indeed implements separate societies with small information exchange.

Position update and boundary handling are the same as in standard PSO with just one more index. Indeed, in a implementation restoring to a vector-capable language, like Matlab, update rules are still a one-line command.

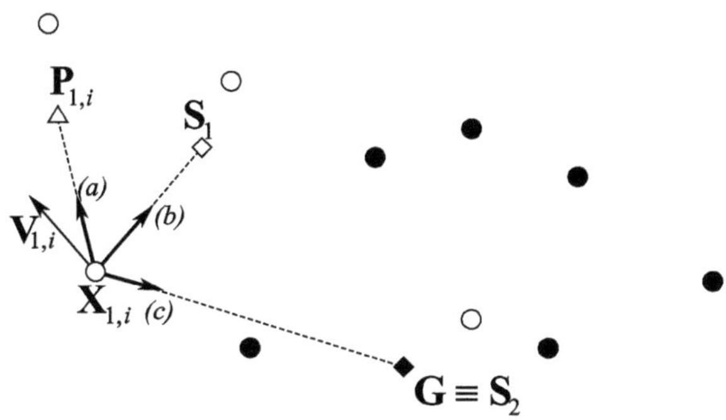

Fig. 6.3. Meta-PSO basic layout. Forces over a generic particle j of swarm 1: (a) pull toward personal best $\mathbf{P}_{1,j}$; (b) pull toward swarm best $\mathbf{S}_1$; (c) pull toward global best $\mathbf{G}$ (here belonging to swarm 2).

6.3.2 Modified Meta-PSO

As an enhancement to MPSO aimed at keeping swarms apart from each other, and hence widening the global search, an inter-swarm repulsion is introduced and a Modified MPSO (or M²PSO) produced. The velocity update rule becomes:

$$\mathbf{V}_{j,i}^{(k+1)} = \omega^{(k)}\mathbf{V}_{j,i}^{(k)} + \phi\eta_1(\mathbf{P}_{j,i} - \mathbf{X}_{j,i}^{(k)}) + \phi\eta_3(\mathbf{S}_j - \mathbf{X}_{j,i}^{(k)}) + \\ + \phi\eta_2(\mathbf{G} - \mathbf{X}_{j,i}^{(k)}) - \sum_{s\neq j} \phi\xi \frac{\mathbf{B}_s^{(k)} - \mathbf{X}_{j,i}^{(k)}}{\left|\mathbf{B}_s^{(k)} - \mathbf{X}_{j,i}^{(k)}\right|^\gamma} \quad (6.6)$$

where the last term is a sum of the repulsions between each single particle and all the *other* swarms barycentra $\mathbf{B}_j^{(k)} = (1/N_p)\sum_{i=1}^{N_p} X_{j,i}^{(k)}$ weighted by a random value ϕ and a fixed weight ξ (Figure 6.4). The repulsive force introduced

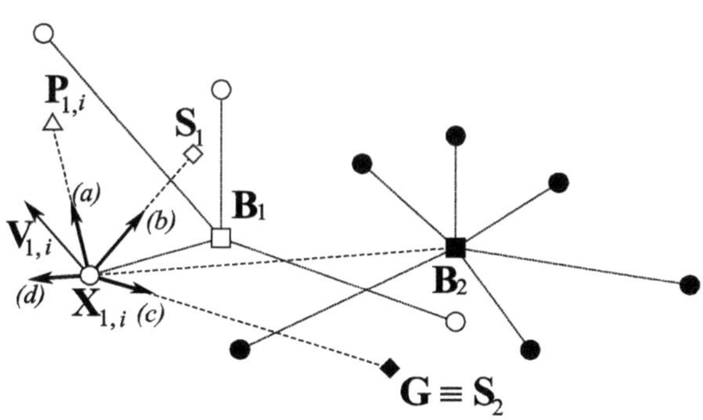

Fig. 6.4. Modified Meta-PSO basic layout. Forces over a generic particle j of swarm 1: (a) pull toward personal best $\mathbf{P}_{1,j}$; (b) pull toward swarm best $\mathbf{S}_1$; (c) pull toward global best $\mathbf{G}$ (here belonging to swarm 2); and (d) inter-swarm repulsion computed from the barycentra of the other swarms.

is a function of distance according to power γ. If $\gamma = 2$, as used here, force decays as the inverse of distance.

6.3.3 Stabilized Modified Meta-PSO

As a further enhancement to the M^{2}PSO it can be ruled that the swarm which is performing best, *i.e.* the swarm j whose social knowledge coincides with the racial knowledge $\mathbf{S}_j = \mathbf{G}$, is not repelled by other swarms, or, in other words, stabilizes itself. This allows for the best swarm to keep exploring the surroundings of the current best position, refining it, whereas other swarms extend the search in other points of the space, hence greatly enhancing the possibility of escaping a local minimum.

6.4 Differentiated Meta-PSO Algorithms

Few variations over the original PSO algorithm were presented in the previous section. In all these algorithms one or more additional pulls (or pushes) were introduced on each particle. These new forces were functions of the knowledge and/or position of the other swarms and acts on all particles of all swarms.

In this section different schemes will be devised, where forces act differently on a particle-by-particle basis.

These *Differentiated* Meta-Swarm algorithm are still based on a multi-swarm approach, but in each swarm a particle is bestowed a special *leader* status and it spans the solution space with an update law that is different from that of the other particles in the same swarm. If the bee similitude often used for PSO is taken, we can think of this special particle as the "queen bee" of each swarm.

Two different flavors of Differentiated Meta-PSO will be presented and discussed here.

6.4.1 Absolute Leader Meta-PSO

In the Absolute Leader Meta-PSO the role played by the leader particle is decided at the beginning and the leader particle never change. Such paradigm will be applied only to the MPSO algorithm, but in principle it can be applied to any Meta-PSO algorithm of the previous Section. Preliminary analyses [18] have shown that the MPSO is the one taking the larger benefit. The resulting algorithms will be denoted shortly as ALMPSO.

In this algorithm the leaders behave indeed as the agents of a MPSO, with an attraction towards the leader personal best (personal knowledge), an attraction towards the swarm best (social knowledge) and an attraction towards the global best of all swarms (racial knowledge). On the other hand all other swarms agents obeys to interactions which are confined within the swarm itself, that is, they are no subject to racial knowledge. The updating rule for the velocity of the MPSO algorithm are therefore modified into:

$$\mathbf{V}_{j,i}^{(k+1)} = \omega^{(k)}\mathbf{V}_{j,i}^{(k)} + \phi\eta_1(\mathbf{P}_{j,i} - \mathbf{X}_{j,i}^{(k)}) + \phi\eta_3(\mathbf{S}_j - \mathbf{X}_{j,i}^{(k)}) + \delta_{j,i}^{L_j}\phi\eta_2(\mathbf{G} - \mathbf{X}_{j,i}^{(k)}) \tag{6.7}$$

while the updating rule for the position remains the usual one:

$$\mathbf{X}_{j,i}^{(k+1)} = \mathbf{X}_{j,i}^{(k)} + \mathbf{V}_{j,i}^{(k+1)} \tag{6.8}$$

In (6.7) $\delta_{ij}^{L_j}$ is a function which value is 1 only if $i = L_j$, being L_j the index denoting the leader of swarm j, otherwise it is 0 (Figure 6.5).

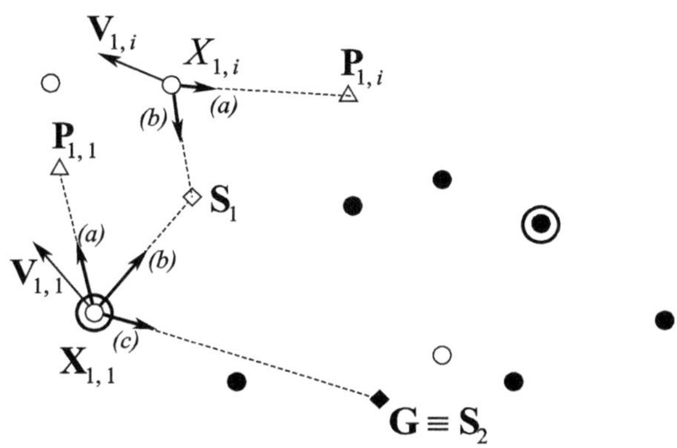

Fig. 6.5. Differentiated Meta-PSO basic layout. Forces over the generic particle are: (a) pull toward personal best $\mathbf{P}_{1,j}$; (b) pull toward swarm best $\mathbf{S}_1$. For what concerns the Leader, he is subject also to (c) pull toward global best $\mathbf{G}$ (belonging to swarm 2).

6.4.2 Democratic Leader Meta-PSO

In the ALMPSO, the choice of the leader within the swarm is indeed arbitrary and made randomly at the beginning. This can be sub-optimum. To devise a different strategy, possibly more performing, leader particle can change within a swarm by setting as leader the particle exhibiting the best performance (the one whose personal best coincides with the swarm's best). This second technique will be indicated as Democratic Leader algorithm in general and, if applied to the Meta-PSO paradigm, as Democratic Leader Meta-PSO (DLMPSO).

Equation (6.7) holds of course also for the Democratic algorithms. Simply, in the Absolute case the value of L_j is chosen at the beginning and never changed (in this case it is computationally simpler to consider $L_j = 1, \forall j = 1\ldots, N_s$); whereas in the Democratic case L_j is updated at each iteration k by having L_j pointing to that agent for which $\mathbf{P}_{j,i} = \mathbf{S}_j$. In both cases the algorithm complexity is not significantly different than that of a MPSO.

6.5 Algorithm Tuning

To asses the optimization capabilities of the algorithm a first set of optimizations over standard test functions have been performed, in a single-objective framework.

These results have been published in [3, 18, 32] where a sinc-like test function was introduced and exploited:

$$\mathscr{F}(\mathbf{X}) = 1 - \prod_{i=1}^{N} \frac{\sin[\pi(x_i - q_i)]}{\pi(x_i - q_i)} \tag{6.9}$$

being $\mathbf{X} = [x_1, x_2, \ldots, x_N]$ a point in the N-dimensional solution space and $\mathbf{Q} = [q_1, q_2, \ldots, q_N]$ the predefined global minimum of the cost function. This function has been chosen since it presents several local minima and it is particularly challenging since the global minimum is a single point whereas local minima extends along lines, hence making the probability that a local minima is hit much higher.

Other test functions investigated were the Ackley function [24]:

$$\mathscr{F}(\mathbf{X}) = 20 + e - 20e^{-0.2\sqrt{\frac{1}{N}\sum_i x_i^2}} - e^{\frac{1}{N}\sum_i \cos(2\pi x_i)} \tag{6.10}$$

the Rosenbrock function [24]:

$$\mathscr{F}(\mathbf{X}) = \sum_{i=1}^{N} \left(100\left(x_{2i} - x_{2i-1}^2\right)^2 + \left(1 - x_{2i-1}\right)^2\right) \tag{6.11}$$

the Rastrigin function [24]:

$$\mathscr{F}(\mathbf{X}) = \sum_{i=1}^{N} \left(x_i^2 - 10\cos(2\pi \cdot x_i) + 10\right) \tag{6.12}$$

and the Shekel function [33]:

$$\mathscr{F}(\mathbf{X}) = 12 - \sum_{i=1}^{9} \left(\left(\mathbf{X} - \mathbf{a}_i\right)^T \left(\mathbf{X} - \mathbf{a}_i\right) + c_i\right)^{-1} \tag{6.13}$$

In the first three N is the dimension of the domain, while, in the last one the domain is bi-dimensional, $\mathbf{X} = (x_1, x_2)$ and, a_i are the vectors of the i-th local minimum and c_i are constant proportional to the minimum value $\mathscr{F}\left((\mathbf{a}_i)^T\right) \approx 12 - \frac{1}{c_i}$. In this investigation there are 9 minima at points $(-3, -3)$, $(-3, 0)$, $(-3, +3)$ ... $(+3, +3)$.

The application of the Meta-PSO schemes to these test functions were mainly intended to demonstrate the reliability of the Meta-Swarm approach, especially as compared to the conventional PSO. Figure 6.6, elaborated after

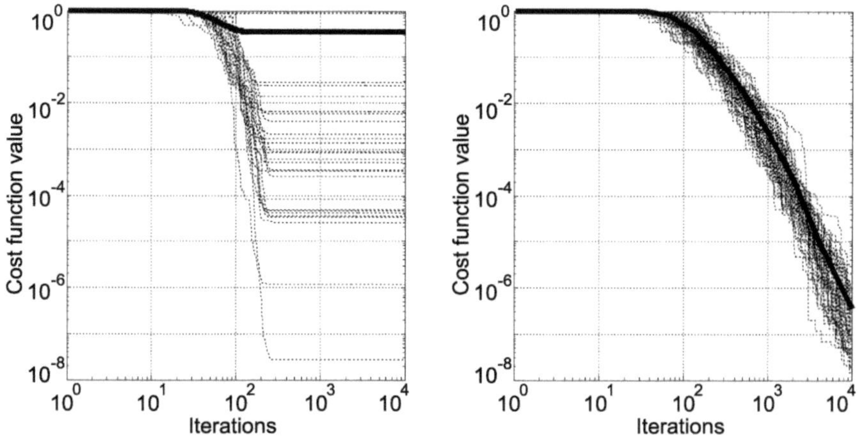

Fig. 6.6. Best agent behavior over 50 independent trials for PSO (left) and SM^{2}PSO (right). Dashed lines are single trials, thick darker line is average over trials.

the data presented in [3] proves this. Light blue curves are 50 independent PSO optimizations over a $N = 10$ sinc-like function (6.9) with $q_i = 3\ \forall i = 1, \ldots, 10$ lasting 10000 iterations. It is worth noticing how PSO gets trapped in local minima and therefore nearly half of the trials gives very poor cost function values, pretty close to 1. The average over these 50 trials is represented by the thick dark blue lines.

On the other hand the SM^{2}PSO algorithm, in red, shows how different trials (pale red) steadily converges toward the minimum without being trapped in local minima and hence its average results (dark, thick red line) is much better. It is worth noticing that both PSO *might* reach the global minimum, but then Meta-PSOs attain this global minimum nearly always, and hence, in this sense, they are more reliable.

The key point of Meta-PSO algorithms is of course, as in any other stochastic optimization algorithm, the correct selection of the optimization parameters, here the η_i pulls weights and the ξ repulsion weight.

An extensive study of the convergence of the different algorithms as these weights are let to vary has been presented in [18]. The results presented in the aforementioned paper are summarized in Tab. 6.1. These results were obtained over the $N = 10$ sinc-like function (6.9) and by maintaining $\eta_1 = 2$ as it is common practice in PSO, and by varying the other parameters in the $[0, 3]$ range.

Furthermore, these results were obtained with a number of swarms equal to 4, and a number of particles within each swarm equal to 18. Optimizations were performed over 2000 iterations and 20 separate, independent, optimization were carried out for each case to provide an average behavior. The results might very well be non-optimal since they are attained on a particular test function and a particular set up, but nonetheless they present a good assessment, and a further investigation starting point.

Table 6.1. Optimal weights

Algorithm	η_1	η_2	η_3	ξ
MPSO	2	$(\eta_2 + \eta_3 > 4)$		N/A
M^2PSO	2	2	> 2	$\simeq 0.5$
SM^2PSO	2	2	> 1.5	< 1.5
ALMPSO	2	< 2.5	> 2	N/A
DLMPSO	2	< 2.5	2.5	N/A

As a general remark the weights η_i should maintain a value equal to 2 for all the Meta-swarm algorithms, whereas ξ, representing the force pushing swarms away must be lower. The optimal value is higher for the SM^2PSO since the stabilization factor prevents the best swarm from being repelled from the global best.

For what concerns the other test functions (6.10)-(6.13), the different methods have been compared among them and with the standard PSO in terms of average cost attained and standard deviation over a suitable average of runs. Parameters were chosen according to Tab. 6.1 and are $\eta_1 = 2$, $\eta_2 = 2$, $\eta_3 = 2$, $\xi = 2/3$, $N_s = 4$, $N_p = 20$. Results are presented in Tab. 6.2 (From [18]) where the last row shows the dimension N of the parameter space, the number of

Table 6.2. Performances with different cost functions: average final value and, in parenthesis, the standard deviation (From [18])

	Ackley	Rosenbrock	Rastrigin	Sheckel
PSO	3.149	0.218	30.062	0.0022
	(0.768)	(0.039)	(9.485)	(0.0178)
MPSO	3.638	0.119	22.493	0.0044
	(0.734)	(0.042)	(8.019)	(0.0250)
M^2PSO	1.358	0.812	16.697	0.0017
	(0.860)	(1.173)	(8.819)	(0.0026)
SM^2PSO	1.935	0.481	23.702	0.0026
	(0.779)	(0.138)	(8.118)	(0.0178)
ALMPSO	3.643	0.748	19.545	0.0004
	(0.701)	(0.236)	(6.488)	(0.0005)
DLMPSO	3.523	0.640	24.378	0.0004
	(0.583)	(0.202)	(7.663)	(0.0004)
N	30	30	30	2
iterations	10^5	10^5	10^5	200
samples	27	28	28	66

iterations of the algorithms and the number of samples, that is the number of separate independent runs on which the averages are computed. It is apparent that no one of the proposed scheme is absolutely the best, but depending on the considered test function one algorithm could perform better than the other; in any case, there is always at least one Meta-PSO performing better than the standard PSO.

6.6 Meta-PSO for Multi-Objective Optimization

In order to extend the Meta-PSOs to multi-objective problems, we choose to adopt the previous introduced directed multi-objective scheme, since it allows to handle more than two objectives, even if the intrinsic multi swarm structure of the Meta-PSOs could be used also in conjunction with the VEPSO.

In the directed multi-objective (dMO) PSO, the personal best is selected as the first non-dominated solution explored by the agent. In Meta-PSOs, at each iteration the personal, the social and the racial knowledge have to be updated. Dealing with the dMO scheme, it means that at each step non-dominated solutions are dynamically updated and stored. When a personal, social or global best solution is found to be dominated by a new solution, the new dominant solution overwrite the dominated one. Therefore the presence of different swarms is useful because several non-dominated solutions can be taken into account and attract at different levels (personal, social and global) the particles that are searching the solution space.

In order to test the Meta-swarms schemes in multi-objective optimization, they have been applied to the optimization of the two bi-dimensional functions

$$\begin{cases} f_1(\mathbf{x}) = x_1 \\ f_2(\mathbf{x}) = \frac{1+x_2}{x_1} \end{cases} \quad \text{where:} \begin{cases} x_2 + 9x_1 \geq 6 \\ -x_2 + 9x_1 \geq 1 \end{cases} \tag{6.14}$$

and

$$\begin{cases} g_1(\mathbf{x}) = x_1 \\ g_2(\mathbf{x}) = x_2 \end{cases} \quad \text{where:} \begin{cases} x_1^2 + x_2^2 - 1 - \frac{1}{10}\cos\left(16\tan^{-1}\frac{x_1}{x_2}\right) \geq 0 \\ (x_1 - 0.5)^2 + (x_2 - 0.5)^2 \leq 0.5 \end{cases} \tag{6.15}$$

The choice of these two-objective problems presents the advantage that with them it is possible to graphically represent the Pareto front and therefore to control how the different algorithms work in the definition of it. As in the single-objective case, also here we compare among them and with the standard PSO the different Meta-PSOs, limiting however our analysis to the three (the MPSO and the two differentiated schemes derived from it) that have shown the best performances in the single-objective optimization. Here the comparison in carried out in terms of capability in determining the Pareto front and of number of dominant solutions found by each scheme.

Figure 6.7 reports the results relative to the problem defined by (6.14). In light gray it is represented the domain of definition of the function, while the

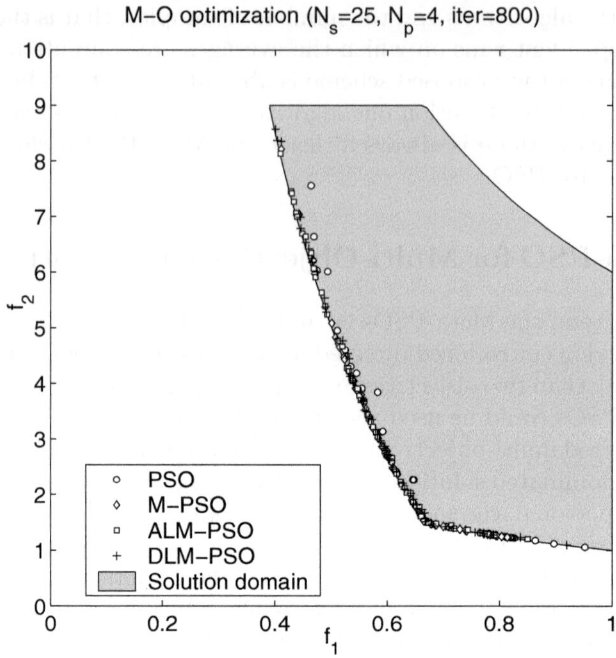

Fig. 6.7. Domain of multi-objective problem defined in (6.14) and distribution of dominant solutions found by different optimization approaches after 800 iterations.

different curves accumulating at its boundary represent the front of dominant solutions determined with the standard PSO and the three Meta-PSOs. From this figure it appears that Meta-PSO variations are more efficient and reliable in this task than traditional PSO. In fact, dominant solutions found by Meta-PSO are closer to the actual Pareto front than solutions found by PSO. Moreover, the number of dominant solutions found by Meta-PSO is higher than those found by PSO for the same number of iterations. This is essentially due to the fact that PSO and similar population-based are essentially aimed at finding one best, even if several techniques can be exploited to extend their solution diversity capabilities. Meta-PSO, on the other hand, are inherently able to diversify, thanks to the several Swarm best kept track of. This leads to an implicit parallelism in searching the Pareto-solutions which is more efficient than a mere repeated application of methods which do not implement a parallel approach.

In Figure 6.8, similar results, obtained for the problem defined by (6.15), are shown. Also in this particular case, in which the Pareto front is not continuous, the Meta-PSOs outperform the standard PSO.

Figure 6.9 reports the evolution of the front of dominant solutions that have been found at different iteration number, by the different techniques considered here, for the function (6.14). Although PSO seem to be very fast in finding

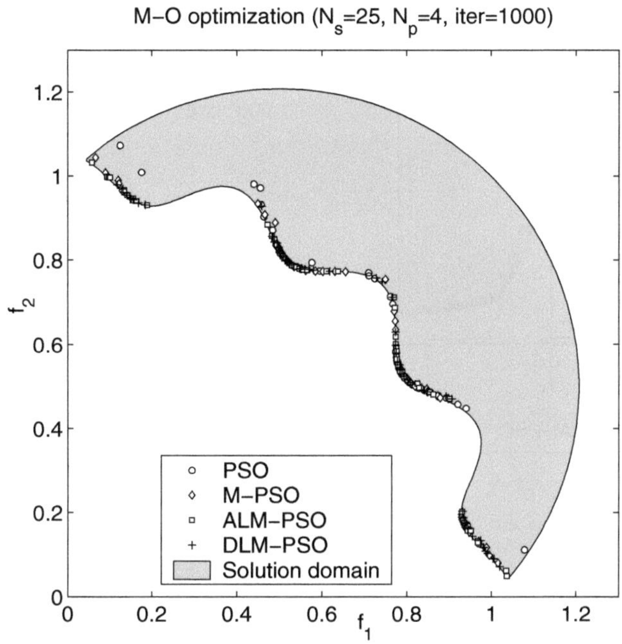

Fig. 6.8. Domain of multi-objective problem defined in (6.15) and distribution of dominant solutions found by different optimization approaches after 1000 iterations.

Table 6.3. Number of dominant solutions for function (6.14) found by different optimization approaches after N_{iter} iterations.

N_{iter}	40	80	200	800
PSO	28	28	31	39
MPSO	25	36	41	62
ALMPSO	28	36	41	62
DLMPSO	34	34	45	63

solutions early in the run, the Meta-PSO techniques improve the search during the optimization process. This is also evident from the results summarized in Table 6.3, in which it is reported the number of dominant solutions at the different steps shown in Figure 6.9. While increasing the number of iteration the number of dominant solution found by the standard PSO remains almost constant, for the other three approaches it continue to rise, in almost the same way. The same considerations apply for Figure 6.10 and Table 6.4, referred to function (6.15).

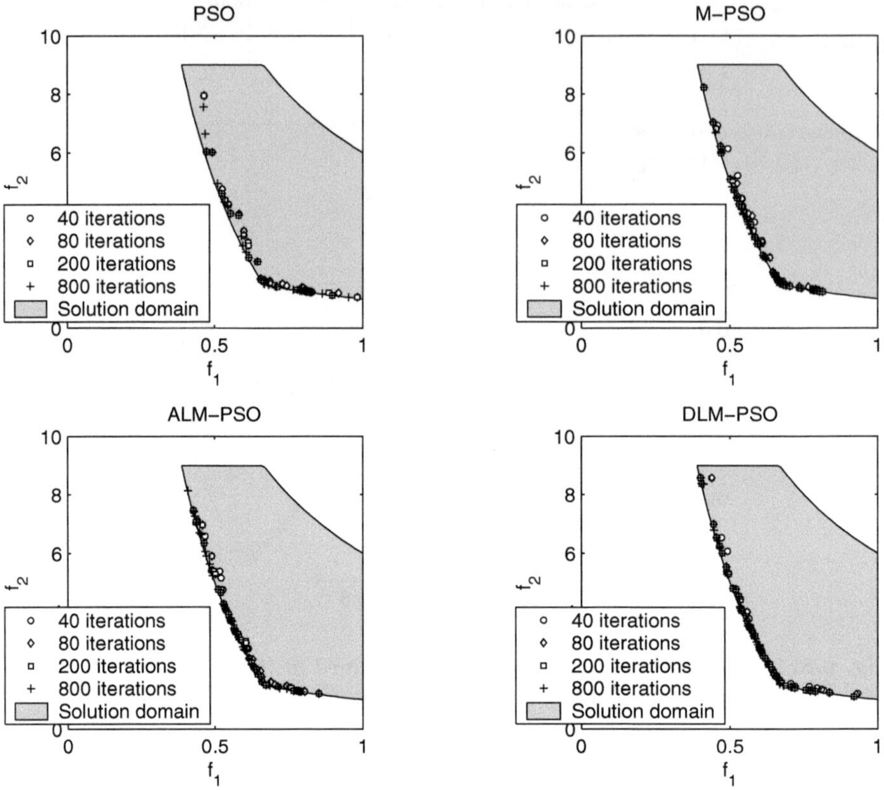

Fig. 6.9. Dominant solutions for function (6.14) found at different iteration numbers by the considered optimization approaches.

Table 6.4. Number of dominant solutions for function (6.15) found by different optimization approaches after N_{iter} iterations.

N_{iter}	50	100	250	1000
PSO	30	37	43	49
MPSO	25	31	44	64
ALMPSO	30	33	41	67
DLMPSO	22	27	42	63

From the preliminary analysis presented so far, it is clear that the Meta-PSO techniques are able to properly handle multi-objective optimization problems and they also outperform the classical Particle Swarm method, both in terms of number of dominant solutions found during the optimization process and of their capability to reach the Pareto front.

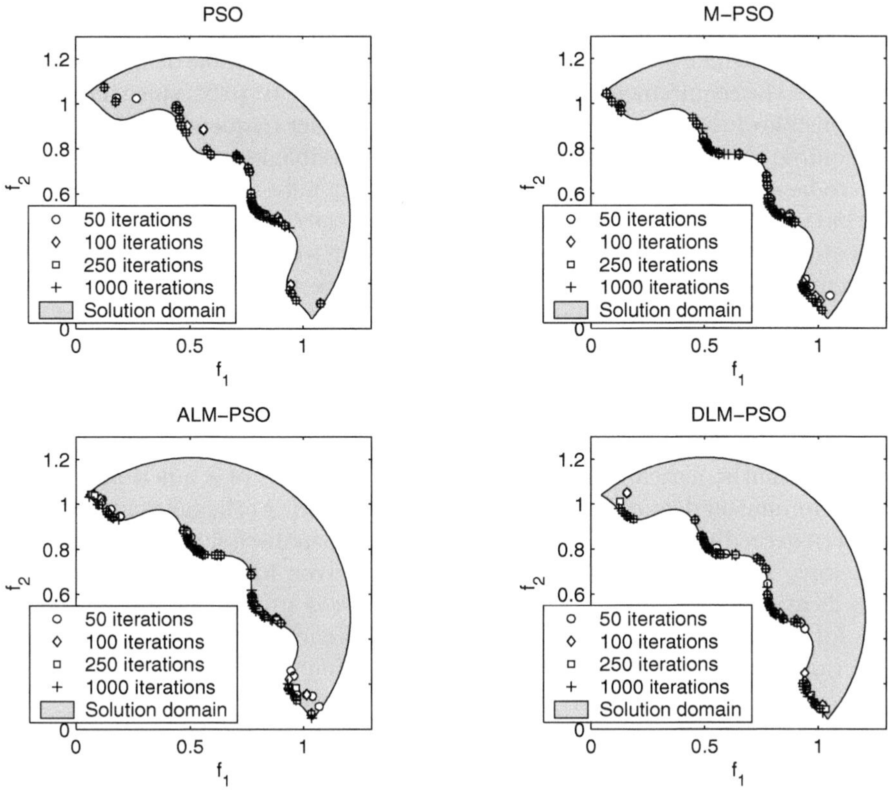

Fig. 6.10. Dominant solutions for function (6.15) found at different iteration numbers by the considered optimization approaches.

6.7 Dual Band Linear Array Design

In order to show the Meta-PSO algorithms effectiveness against multi-objective engineering problems and to extend their range of applicability, we have used the three schemes analyzed in the previous section for the design optimization of an antenna array. In the design of an antenna array several constraints have to be satisfied: some of them could be non-linear, and this makes them suitable to be optimized via an evolutionary approach. Moreover the requirements are often conflicting, both for what concerns the different antenna radiating features, and the possibility of obtaining these without increasing too much the antenna structure. Even if these problems are therefore clearly multi-objective, except for few sporadic cases [34] they are reduced to single objective problems, generally through the WSM methods (see e.g. [18, 35] and reference therein).

Here, we consider the optimization of the array factor of a linear array antenna for a UMTS base station. The antenna consists of 14 dual-band elements whose position and excitation (both amplitude and phase) are free to vary. The constraints on the radiation pattern on both the [1920÷2170] and [2500÷2690] MHz bands, which are relative to UMTS, are reported in Table 6.5: particularly critical is the requirement on the beam scanning from 0° to 8°, since it could be responsible of the onset of grating lobes at the higher frequencies. Moreover, it is required that the antenna has the smallest possible length, that in any case has to be smaller than 1.5 m and that the feeding network is as much as possible balanced. Representing all the constrains on the radiation patterns through one objective function, whose expression is given in the following, and the requirements on the feeding network by another objective function, the design of the "optimum" dual band linear array is reduced to a two-objective problem, that can be faced up with the techniques introduced above.

The first objective function has to take into account that the radiation pattern must fit a specific mask, derivable from the specifications in Table 6.5, in both the UMTS1 and UMTS2 frequency bands and for both the 0° and 8° directions; mathematically, they are expressed by means of a function that has to be minimized defined as the sum of the magnitude of the far field radiation pattern exceeding the prescribed envelope. This penalizes side-lobes above the envelope, while neither penalty nor reward is given for side-lobes below the specification. This kind of constrain is of course non-linear and this is a reason for using an evolutionary optimization approach. The resulting objective function to be minimized is therefore the following:

$$f_1 = \frac{1}{N_\theta} \sum_{\alpha=1}^{A} \sum_{f=1}^{F} \sqrt{\sum_{\theta} err(\theta)_{f,\alpha}^2} \tag{6.16}$$

where $err(\theta)_{f,\alpha}$ are the values of the far field radiation pattern exceeding the mask in point θ at the frequency f and tilt angle α. This means that for each solution 12 radiation patterns are evaluated. It must be noted that the side lobe suppression requirements gives a maximum level for the sidelobes throughout the angular range, while the constraint of the maximum acceptable depth for the first null above the horizon is recasted in requiring the pattern to be above −13 dB from the direction of maximum radiation up to 5° above that direction (as it can be seen in the masks plotted in Figure 6.13). To allow for a single set of position, amplitudes and phases to create multiple beams, the parameters

Table 6.5. Radiation pattern requirements

Electrical down-tilt	Variable between 0° ÷ 8°
Side lobe suppression	> 20 dB between +80° ÷ +90° > 15 dB from main lobe to −30° > 12 dB elsewhere
First null above the horizon	> −13 dB

are used as they are to produce the broadside beam, while, to produce the 8° tilted beam, an appropriate linear phase is added to the elements.

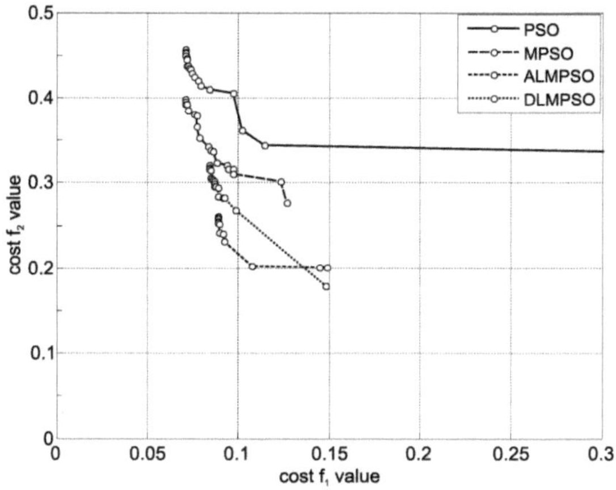

Fig. 6.11. Dominant solutions of multi-objective optimization of the dual band linear array: comparison of the different methods after 1000 iterations.

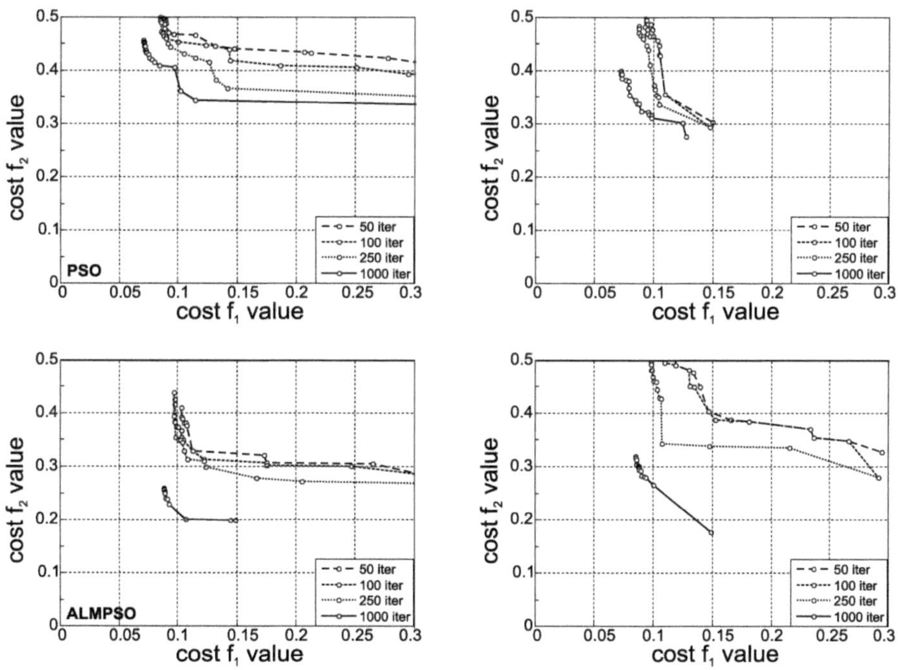

Fig. 6.12. Dominant solutions of multi-objective optimization of the dual band linear array: results found by the considered algorithms at different iteration numbers.

For the optimization of the feeding network, we can consider that the ratio between the minimum and maximum excitation amplitudes a_i^{min}/a_i^{max} must be as close as possible to 1, in order to have as balanced as possible power dividers. Therefore the second objective to be minimized is:

$$f_2 = 1 - \frac{|a_i|^{min}}{|a_i|^{max}} \tag{6.17}$$

The optimization has been carried out acting on both the amplitude and phase of the excitation coefficients and on the position of the array elements and runs on a population composed of 25 swarms of 15 agents, with $\eta_3 = 3$ and $\xi = 0.5$ (where applicable); the results shown in the following have been obtained after 5000 iterations. The radiation pattern of each element of the array has been taken into account through a cosine function.

In Figure 6.11 the Pareto curves obtained with the three considered Meta-PSO schemes are reported, to compare their behavior and their effectiveness with the standard PSO. All the three Meta-swarm approaches works better than the PSO, while among them the worst is the MPSO. It is interesting to notice that, with almost the same value for f_1, the MPSO generates solutions that are more efficient from the point of view of f_2. A similar consideration can be applied to the comparison of the Pareto curves obtained with the two differentiated MPSOs. We can also observe that the MPSO and DLMPSO are the schemes that gives rise to the greater dynamic among results; in fact also ALMPSO, even if less remarkably than the PSO, is inclined to generate solutions in which only f_2 changes, while f_1 remains almost constant.

In Figure 6.12 it is shown how the the Pareto curves obtained with the different schemes evolve during the optimization process: while the variation of the Pareto curves is not so noticeable with the MPSO, using the differentiated Meta-swarms approaches one starts from sets of solution that are more spread out, but that converges to better values in correspondence of further steps of the procedure.

In Figures 6.13 and 6.14 sample results of the radiation patterns of the optimized antenna are reported; they are computed at the extremes and at the central frequencies of the two considered sub-bands (top and bottom), for the cases of broadside maximum radiation (Figure 6.13) and of the maximum beam tilt (Figure 6.14): in all cases the antenna constrains, visualized by the mask shown on the plots, have been satisfied.

The values of the parameters that give these results are reported in Table 6.6: from the first column, in which it is listed the position of the elements it appears that the total length of the antenna is about 1.3 m. The others two columns show the amplitude and the phase of the excitation coefficients.

The results reported show that the considered Meta-PSO techniques have a superior capability than standard PSO on multi-objective optimizations of EM structures. In particular, multi-objective implementations of the Meta-PSO algorithms are able to properly handle different design criteria and to explore the Pareto front more efficiently than standard PSO. Finally, the resulting

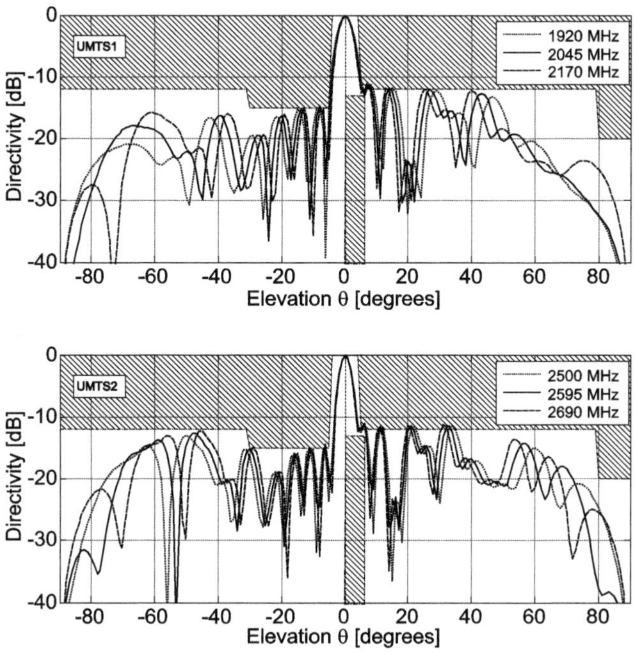

Fig. 6.13. Resulting radiation pattern for the optimized beam scanning linear array in no tilt configuration; lower (top) and higher band response (bottom). (from [35])

Table 6.6. Optimized distribution of excitation and position of each element for the linear array.

N.	Position (cm)	$\|a_i\|$	phase (deg)
1	0.00	0.96	71.63
2	9.66	0.62	44.11
3	21.49	0.68	125.66
4	27.81	0.64	39.31
5	43.91	0.68	65.55
6	50.48	0.57	137.83
7	60.46	0.94	77.80
8	69.54	0.81	76.64
9	82.30	0.81	92.32
10	93.22	0.85	71.54
11	100.94	0.58	98.54
12	111.46	0.96	73.37
13	123.30	1.00	71.82
14	134.00	0.92	53.61

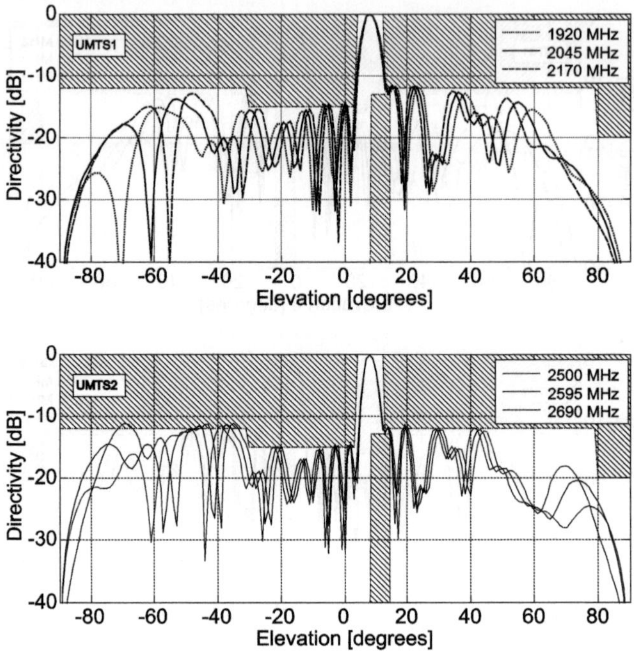

Fig. 6.14. Resulting radiation pattern for the optimized beam scanning linear array in maximum tilt configuration; lower (top) and higher band response (bottom). (from [35])

optimized antennas have improved performance over periodic arrays and previously presented aperiodic arrays [35], especially under the point of view of the power divider balance.

6.8 Conclusions

Several variations over the PSO algorithm exploiting multiple swarms and different velocity update strategies have been presented. The performances of the proposed techniques with respect to the standard PSO algorithm have been assessed both on single and multi-objective analytical test functions and a good range for the algorithm weights was given.

The multi-objective Meta-PSO engine has then been used to design dual band antenna arrays as an example for its applications in engineering EM problems. The proposed techniques present very good optimization performances over the multi-objective problem of the optimization of antenna arrays and prove to outperform conventional PSO.

References

1. Fonseca, C.M., Fleming, P.J.: Multiobjective Optimization and Multiple Constraint Handling with Evolutionary Algorithms Part I and II. IEEE Trans. Syst., Man, Cybern. 28(1), 26–37, 38-47 (1997)
2. Miettinen, K.: Nonlinear Multiobjective Optimization. Kluwer Academic Publishers, Boston (1999)
3. Selleri, S., Mussetta, M., Pirinoli, P., Zich, R.E., Matekovits, L.: Some Insight over New Variations of the Particle Swarm Optimization Method. IEEE Antennas and Wireless Propagation Letters 5, 235–238
4. Reyes-Sierra, M., Coello Coello, C.A.: Multi-Objective Particle Swarm Optimizers: A Survey of the State-of-the-Art. Int. Journal of Computational Intelligence Research 2(3), 287–308 (2006)
5. Mostaghim, S., Teich, J.: Covering Pareto-optimal Fronts by Subswarms in Multi-objective Particle Swarm Optimization. In: Proc. of Congress on Evolutionary Computation, CEC 2004, vol. 2, June 2004, pp. 1404–1411 (2004)
6. Eberhart, R.C., Shi, Y.: Comparison between genetic algorithms and particle swarm optimization. In: Proc. of 7th Annual Conf. Evol. Program., March 1998, pp. 611–616 (1998)
7. Hodgson, R.J.W.: Particle swarm optimization applied to the atomic cluster optimization problem. In: Proc. of Genetic and Evolut. Comput. Conf., pp. 68–73 (2002)
8. Robinson, J., Rahmat-Samii, Y.: Particle swarm optimization in electromagnetics. IEEE Trans. Antennas Propagat. 52, 397–407 (2004)
9. Boeringer, D.W., Werner, D.H.: Particle swarm optimization versus genetic algorithms for phased array synthesis. IEEE Trans. Antennas Propagat. 52, 771–779 (2004)
10. Matekovits, L., Mussetta, M., Pirinoli, P., Selleri, S., Zich, R.E.: Particle swarm optimization of microwave microstrip filters. In: 2004 IEEE AP-S Symposium Digests, Monterey (CA), June 20-26 (2004)
11. Gies, D., Rahmat-Samii, Y.: Reconfigurable array design using parallel particle swarm optimization. In: IEEE AP-S Symposium Digests, June 22-27, 2003, pp. 177–180 (2003)
12. Ciuprina, G., Ioan, D., Munteanu, I.: Use of intelligent-particle swarm optimization in electromagnetics. IEEE Trans. on Magnetics 38, 1037–1040 (2002)
13. Matekovits, L., Mussetta, M., Pirinoli, P., Selleri, S., Zich, R.E.: Improved PSO Algorithms for Electromagnetic Optimization. In: 2005 IEEE AP-S Symposium Digests, Washington (DC), July 3-8 (2005)
14. Jin, N., Rahmat-Samii, Y.: Parallel particle swarm optimization and finite-difference time-domain (PSO/FDTD) algorithm for multiband and wide-band patch antenna designs. IEEE Trans. Antennas Propagat. 53, 3459–3468 (2005)
15. Cui, S., Weile, D.S.: Application of a parallel particle swarm optimization scheme to the design of electromagnetic absorbers. IEEE Trans. Antennas Propagat. 53, 3616–3624 (2005)
16. Jin, N., Rahmat-Samii, Y.: IEEE Advances in Particle Swarm Optimization for Antenna Designs: Real-Number, Binary, Single-Objective and Multiobjective Implementations. IEEE Trans. Antennas Propagat. 55, 556–567 (2007)
17. Moradi, A., Fotuhi-Firuzabad, M.: Optimal Switch Placement in Distribution Systems Using Trinary Particle Swarm Optimization Algorithm. IEEE Trans. Power Deliv. 23, 271–279 (2008)

18. Selleri, S., Mussetta, M., Pirinoli, P., Zich, R.E., Matekovits, L.: Differentiated Meta-PSO Methods for Array Optimozation. IEEE Trans. Antennas Propagat. 56 (January 2008)
19. Parsopoulos, K., Vrahatis, M.: Recent approaches to global optimization problems through particle swarm optimization. Natural Computing 1, 235–306 (2002)
20. Velduizen, D., Zydallis, J., Lamont, G.: Considerations in engineering parallel multiobjective evolutionary optimizations. IEEE Trans. Evol. Comput. 7(2), 144–173 (2003)
21. Genovesi, S., Monorchio, A., Mittra, R., Manara, G.: A Sub-boundary Approach for Enhanced Particle Swarm Optimization and its Application to the Design of Artificial Magnetic Conductors. IEEE Trans. Antennas Propagat. 55, 766–770 (2007)
22. Kennedy, J.: Stereotyping: improving particle swarm performance with cluster analysis. In: Proc. of Congress on Evolutionary Computation, Washington DC, July 6-9, vol. 3, pp. 1931–1938 (1999)
23. Shi, Y., Krohling, R.A.: Co-evolutionary particle swarm optimization to solve min-max problems. In: Proc. of Congress on Evolutionary Computation, Honolulu, HI, May 12-17, vol. 2, pp. 1682–1687 (2002)
24. van den Bergh, F., Engelbrecht, A.P.: A cooperative approach to particle swarm optimization. IEEE Tans. Evol. Comput. 8, 225–239 (2004)
25. Kennedy, J.: Small worlds and mega-minds: effect of neighborhood topology on particle swarm performance. In: Proc. of Congress on Evolutionary Computation, Washington DC, July 6-9, vol. 3, pp. 1931–1938 (1999)
26. Kennedy, J., Eberhart, R.C.: Swarm Intelligence. Morgan Kaufmann, San Francisco (2001)
27. Eberhart, R.C., Shi, Y.: Particle swarm optimisation: developments, applications and resources. In: Proc. of Congress on Evolutionary Computation, pp. 81–86 (2001)
28. Bajpai, P., Singh, S.N.: Fuzzy Adaptive Particle Swarm Optimization for Bidding Strategy in Uniform Price Spot Market. IEEE Trans. Power Syst. 22, 2152–2160 (2007)
29. Xu, S., Rahmat-Samii, Y.: Boundary Conditions in Particle Swarm Optimization Revisited. IEEE Trans. Antennas Propagat. 55, 760–765 (2007)
30. Mikki, S.M., Kishk, A.A.: Hybrid Periodic Boundary Condition for Particle Swarm Optimization. IEEE Trans. Antennas Propagat. 55, 3251–3256 (2007)
31. Mansour, M.M., Mekhamer, S.F., El-Sherif El-Kharbawe, N.: A Modified Particle Swarm Optimizer for the Coordination of Directional Overcurrent Relays. IEEE Trans. Power Deliv. 22, 1400–1410 (2007)
32. Mussetta, M., Selleri, S., Pirinoli, P., Zich, R., Matekovits, L.: Improved Particle Swarm Optimization algorithms for electromagnetic optimization. Journal of Intelligent and Fuzzy Systems 19, 75–84 (2008)
33. Olcan, D.I., Kolundzija, B.M.: Adaptive random search for antenna optimization. In: IEEE Proc. Antennas and Propagation Society International Symposium, June 2004, vol. 1, pp. 1114–1117 (2004)
34. Jin, N., Rahmat-Samii, Y.: Advances in Particle Swarm Optimization for Antenna Designs: Real-Number, Binary, Single-Objective and Multiobjective Implementations. IEEE Trans. Antennas and Propagation 5(3), 556–567 (2007)
35. Mussetta, M., Pirinoli, P., Selleri, S., Zich, R.E.: Differentiated Meta-PSO Techniques for Antenna Optimization. In: Proc. of ICEAA, Turin, Italy, September 2007, vol. 53, pp. 2674–2679 (2007)

7

Multi-Objective Wavelet-Based Pixel-Level Image Fusion Using Multi-Objective Constriction Particle Swarm Optimization

Yifeng Niu[1], Lincheng Shen[1], Xiaohua Huo[2], and Guangxia Liang[3]

[1] College of Mechatronic Engineering and Automation, National University of Defense Technology, Changsha, 410073, China
{niuyifeng,lcshen}@nudt.edu.cn

[2] Equipment Academy of Air Force, Beijing, 100085, China
cindy.huo@gmail.com

[3] College of Mathematics and Computer Science, Hunan Normal University, Changsha, 410081, China
lgx025@163.com

In most methods of pixel-level image fusion, determining how to build the fusion model is usually based on people's experience, and the configuration of fusion parameters is somewhat arbitrary. In this chapter, a novel method of multi-objective pixel-level image fusion is presented, which can overcome the limitations of conventional methods, simplify the fusion model, and achieve the optimal fusion metrics. First the uniform model of pixel-level image fusion based on discrete wavelet transform is established, two fusion rules are designed; then the proper evaluation metrics of pixel-level image fusion are given, new conditional mutual information is proposed, which can avoid the information overloaded; finally the fusion parameters are selected as the decision variables and the multi-objective constriction particle swarm optimization (MOCPSO) is proposed and used to search the optimal fusion parameters. MOCPSO not only uses mutation operator to avoid earlier convergence, but also uses a new crowding operator to improve the distribution of nondominated solutions along the Pareto front, and introduces the uniform design to obtain the optimal parameter combination. The experiments of MOCPSO test, multi-focus image fusion, blind image fusion, multi-resolution image fusion, and color image fusion are conducted. Experimental results indicate that MOCPSO has a higher convergence speed and better exploratory capabilities than MOPSO, especially when the number of objectives is large, and that the fusion method based on MOCPSO is is suitable for many types of pixel-level image fusion and can realize the Pareto optimal image fusion.

N. Nedjah et al. (Eds.): Multi-Objective Swarm Intelligent Systems, SCI 261, pp. 151–178.
springerlink.com

7.1 Introduction

Image fusion is a valuable process in combining images with various spatial, spectral and temporal resolutions to form new images with more information than that can be derived from each of the source images for improving the performances of the fused images in information content, resolution, and reliability. Image fusion can be divided into three levels including the pixel level, the feature level and the decision level. Pixel-level image fusion belongs to the lowest level and can save most original information [1]. The multi-resolution method of pixel-level image fusion is mostly used, the performance of the method based on discrete wavelet transform (DWT) is better than others[2], [3], [4], [5], [6].

Different methods of image fusion have the same objective, i.e. to acquire a better fusion effect. Different methods have the given parameters, and the different parameters may gain different fusion effects. In general, we give the parameters based on the experience or the image contents, so it is fairly difficult to gain the optimal fusion effects. If one image is regarded as one information dimension or a feature subspace, image fusion can be regarded as an optimization problem in several information dimensions or the feature space. The better result, even the optimal result, can be acquired through searching the optimal parameters during the course of image fusion. Therefore, both the proper search objectives and strategy are important for the optimization problem. Nasrabadi[7] primarily explored the problem, and introduced simulated annealing (SA) into the optimization of image fusion. Qin[8] used particle swarm optimization (PSO) to search the fusion weight. However, their single objective for optimization can't meet the real demands, and the algorithm, such as PSO and SA, are relatively simple.

In fact, the evaluation metrics of pixel-level image fusion can be regarded as the optimization objectives. However, these metrics are various, and the different metrics may be are compatible or incompatible with one another, so a good evaluation metric system of pixel-level image fusion must balance the advantages of the diverse metrics. The conventional solution is to change the multi-objective optimization problem into a single objective optimization problem using a weighted linear method. However, the relation of the metrics is often nonlinear, and this method needs to know the weights of different metrics in advance. So it is highly necessary to introduce multi-objective optimization methods based on the Pareto theory to search the optimal parameters in order to realize the optimal pixel-level image fusion. We have proposed a method of multi-objective pixel-level image fusion in [9], and got some meaningful results, while it is primary, moreover, the evaluation metrics need to be improved. There exists information overlapping in the metrics, so we propose new metrics of conditional mutual information which can effectively avoid information overlapping In this chapter. We also try to expand the method of multi-objective image fusion to multi-focus image fusion, multi-resolution image fusion, color image fusion, and even blind image fusion etc.

At present, the representative multi-objective optimization algorithms include PASE (Pareto Archive Evolutionary Strategy)[10], SPEA2 (Strength Pareto Evolutionary Algorithm 2)[11], NSGA-II (Nondominated Sorting Genetic Algorithm II)[12], NSPSO (Non-dominated Sorting Particle Swarm Optimization)[13], MOPSO (Multiple Objective Particle Swarm Optimization) [14],[15], etc. Ref. [14] and [16] conclude that MOPSO has a higher convergence speed and better optimization capacities than other algorithms. However, MOPSO uses an adaptive grid [10] to record the searched particles, and the grid number recommended to divide the objective space is thirty. The grid needs a great quantity of storage about n^{30} where n is the number of objectives. Once n is a little greater, the quantity will need too much calculation time, and cause failure in allocating memory even in integer format. On the other hand, if the number is too small, MOPSO will not embody its superiority in searching. On condition that n is not smaller than 3, MOPSO will show its ability unequal to its ambition.

Using MOPSO and NSGA-II for reference, we presented multi-objective constriction particle swarm optimization (MOCPSO) [17]. In MOCPSO, we do not use the adaptive grid, design a new crowding distance to maintain the population diversity, use an adaptive mutation operator to improve the search capacities and avoid the earlier convergence, and use the uniform design to obtain the optimal parameter combination. In this study, MOCPSO is introduced and applied to optimize the parameters of multi-objective pixel-level image fusion in order to realize the Pareto optimal image fusion. Experiments show that MOCPSO has a higher convergence speed and better exploratory capabilities than MOPSO and NSGA-II, and that the method of multi-objective pixel-level image fusion based on MOCPSO is fairly successful.

The remainder of this chapter is organized as follows. The fundamentals of discrete wavelet transform (DWT) are presented in Section 7.2. The methodology of multi-objective pixel-level image fusion based on DWT is designed in Section 7.3. The evaluation metrics of pixel-level image fusion are established in Section 7.4. The algorithm of multi-objective constriction particle swarm optimization (MOCPSO) is designed in Section 7.5. The experimental results and analysis are given in Section 7.6. Finally, a review of the results and the future research areas are discussed in Section 7.7.

7.2 Fundamentals of Wavelet Transform

The fundamental idea behind wavelet theory is the decomposition of signals into components at different scales or resolutions[18]. The advantage of this decomposition is that signal trends at different scales can be isolated and studied. Global trends can be examined at coarser scales while local variations are better analyzed at fine scales.

The wavelet analysis procedure is to adopt a wavelet prototype function, called an "mother function", or "analyzing wavelet". Temporal analysis is performed with a contracted, high-frequency version of the prototype wavelet,

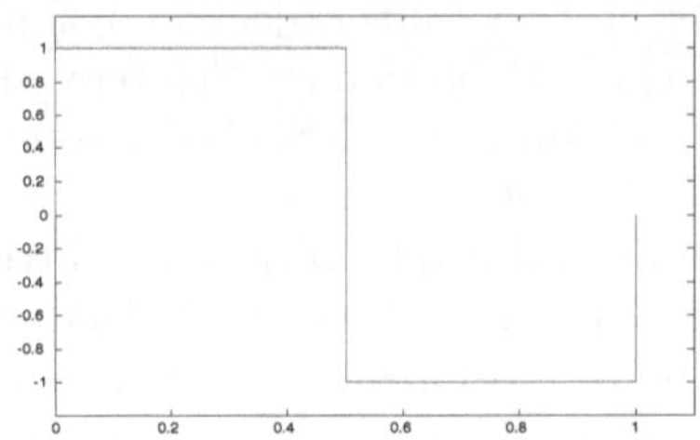

Fig. 7.1. Haar wavelet $\psi(x)$

while frequency analysis is performed with a dilated, low-frequency version of the same wavelet. Because the original signal can be represented in terms of a wavelet expansion (using coefficients in a linear combination of the wavelet functions), data operations can be performed using just the corresponding wavelet coefficients.

The different mother wavelets make different trade-off between how compactly the basis functions are localized in space and how smooth they are. However, Once we know about the mother functions, we know everything about the basis. The Haar wavelet is even simpler, shown in Fig. 7.1, and it is often used in many applications. The Haar scaling function is defined as

$$\phi(x) = \begin{cases} 1, 0 \le x < 1 \\ 0, \text{otherwise} \end{cases} \tag{7.1}$$

The haar wavelet function is defined as

$$\psi(x) = \phi(2x) - \phi(2(x - 1/2)) = \phi(2x) - \phi(2x - 1) \tag{7.2}$$

In order to make more meaningful results, simplify the computation, and avoid the reconstruct distortion, the Haar wavelet is selected the wavelet basis function in this chapter.

7.2.1 Wavelet Transform of 1-D Signals

Dilations and translations of the "mother function", or "analyzing wavelet" $\psi(x)$, define an orthogonal basis, the orthonormal wavelet basis functions:

$$\psi_{j,k}(x) = 2^{j/2}\psi(2^j x - k), k \in Z \tag{7.3}$$

form the wavelet spaces W_j. The variables j and k are integers that scale and dilate the mother function ψ to generate wavelets. The scale index j indicates the wavelet's width, and the location index k gives its position.

Given a multi-resolution analysis $(V_j)_{j \in Z}$ of $L^2(R)$, $\phi(x)$ is a scaling function for V_0, dilations and translations of $\phi(x)$,

$$\phi_{j,k}(x) = 2^{j/2}\phi(2^j x - k) \tag{7.4}$$

form an orthonormal basis for V_j.

Furthermore, the scaling function $\phi(x)$ satisfies the following two-scale dilation equation:

$$\phi(x) = \sqrt{2}\sum_{k} c_k \phi(2x-k) \tag{7.5}$$

for some set of expansion coefficients c_k.

And the wavelet function satisfies the wavelet equation,

$$\psi(x) = \sqrt{2}\sum_{k} d_k \phi(2x-k) \tag{7.6}$$

for some et of expansion coefficients d_k.

Thus, we can expand any function $f(x) \in L^2(R)$ as follows:

$$f(x) = \sum_{k\in Z} a_{J,k}\phi_{J,k}(x) + \sum_{j\geq J}\sum_{k\in Z} b_{j,k}\psi_{j,k}(x) \tag{7.7}$$

where

$$a_{J,k} = \int f(x)\phi_{J,k}(x)dx$$

$$b_{j,k} = \int f(x)\psi_{j,k}(x)dx$$

are the expansion coefficients for $f(x)$.

7.2.2 Wavelet Transform of 2-D Images

Given a multi-resolution analysis $(V_j^{(1)})_{j\in Z}$ of $L^2(R)$, a set of nested subspaces $(V_j^{(2)})_{j\in Z}$ forms a multi-resolution approximation of $L^2(R^2)$ with each vector space $V_j^{(2)}$ being a tensor product of identical 1-D approximation spaces [19]

$$V_j^{(2)} = V_j^{(1)} \otimes V_j^{(1)} \tag{7.8}$$

Furthermore, the scaling function $\Phi(x,y)$ for the 2-D multi-resolution subspaces can be decomposed as

$$\Phi(x,y) = \phi(x)\phi(y) \tag{7.9}$$

where $\phi(x)$ is the 1-D scaling function of the multi-resolution analysis $(V_j^{(1)})_{j\in Z}$. The set of functions

$$\Phi_{j,k,l}(x,y) = \phi_{j,k}(x)\phi_{j,l}(y), j,k,l \in Z \tag{7.10}$$

is an orthonormal basis for $V_j^{(2)}$. Let $\psi(x)$ is the 1-D orthonormal wavelet function correspondingly, the 2-D wavelet subspaces $W_j^{(2)}$ are generated by three wavelet to capture detail information in the horizontal, vertical, diagonal directions

$$\Psi^h(x,y) = \psi(x)\phi(y) \tag{7.11}$$

$$\Psi^v(x,y) = \phi(x)\psi(y) \tag{7.12}$$

$$\Psi^d(x,y) = \psi(x)\psi(y) \tag{7.13}$$

The corresponding orthonormal wavelet basis for $W_j^{(2)}$ is the set

$$\Psi^h_{j,k,l}(x,y) = \psi_{j,l}(x)\phi_{j,k}(y), j,k,l \in Z \tag{7.14}$$

$$\Psi^v_{j,k,l}(x,y) = \phi_{j,l}(x)\psi_{j,k}(y), j,k,l \in Z \tag{7.15}$$

$$\Psi^d_{j,k,l}(x,y) = \psi_{j,k}(x)\psi_{j,l}(y), j,k,l \in Z \tag{7.16}$$

Any image $f(x,y) \in L^2(R^2)$ can be expanded as a sum of its approximate image at some scale J in $V_j^{(2)}$ along with subsequent detail components at scale J and higher.

$$\begin{aligned} f(x,y) = &\sum_{k,l\in Z} a_{J,k,l}\Phi_{J,k,l}(x,y) + \sum_{j\geq J}\sum_{k,l\in Z} b^h_{j,k,l}\Psi^h_{j,k,l}(x,y) + \\ &\sum_{j\geq J}\sum_{k,l\in Z} b^v_{j,k,l}\Psi^v_{j,k,l}(x,y) + \sum_{j\geq J}\sum_{k,l\in Z} b^d_{j,k,l}\Psi^d_{j,k,l}(x,y) \end{aligned} \tag{7.17}$$

with

$$\begin{aligned} a_{J,k,l} &= \int\int f(x,y)\Phi_{J,k,l}(x,y)dxdy \\ b^h_{J,k,l} &= \int\int f(x,y)\Psi^h_{J,k,l}(x,y)dxdy \\ b^v_{J,k,l} &= \int\int f(x,y)\Psi^v_{J,k,l}(x,y)dxdy \\ b^d_{J,k,l} &= \int\int f(x,y)\Psi^d_{J,k,l}(x,y)dxdy \end{aligned}$$

are the wavelet coefficients for $f(x,y)$. The first term on the right band side in (7.17) represents the coarse scale approximation to $f(x,y)$, written as LL_J (horizontal Low-pass, vertical Low-pass); The second term represents the detail component in the horizontal direction, written as $\{LH_j\}$ (horizontal Low-pass, vertical High-pass); The third terms represent the detail component in the vertical direction, written as $\{HL_j\}$ (horizontal High-pass, vertical Low-pass); and the fourth terms represent the detail component in the diagonal direction, written as $\{HH_j\}$ (High-pass in both directions).

The discrete wavelet transform of 2-D image is illustrated as Fig. 7.2.

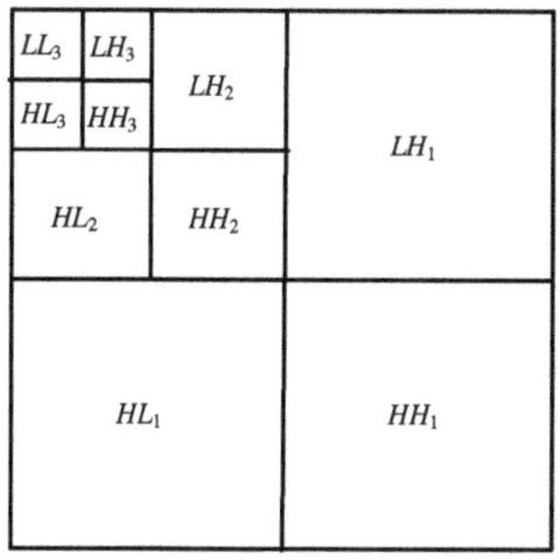

(a) Space-frequency structure of wavelet transform

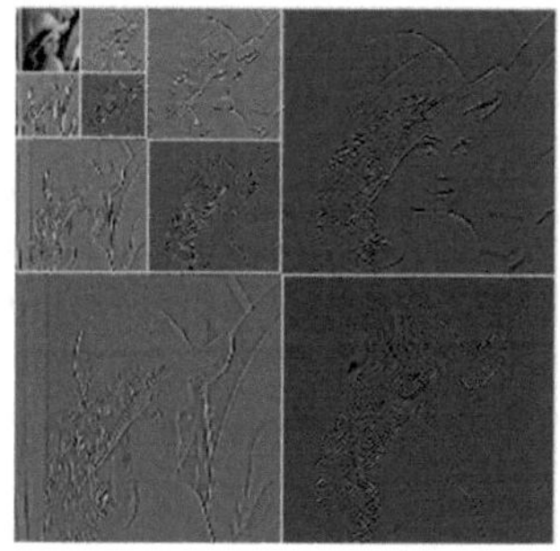

(b) Wavelet transform of the image "Lena"

Fig. 7.2. Illustration of discrete wavelet transform

7.3 Multi-Objective Pixel-Level Image Fusion Based on Discrete Wavelet Transform

As shown in Fig. 7.3, the method of multi-objective pixel-level image fusion based on discrete wavelet transform (DWT) is as follows.

Step 1: Input the registered source images A and B. Find the DWT of each A and B to a specified number of decomposition levels, at each level we will have one-approximation sub band and $3 \times J$ details, where J is the decomposition level. If the value of J is too high, the pixels in sub images will cause the distortion, otherwise the decomposition can't embody the advantage of multiple scales. In general, J is not greater than 3. When J equals zero, the transform result is the original image and the fusion is performed in spatial domain.

Step 2: For the details in DWT domain, salient features in each source image are identified, and have an effect on the fused image [20]. The salient feature is defined as a local energy in the neighborhood of a coefficient.

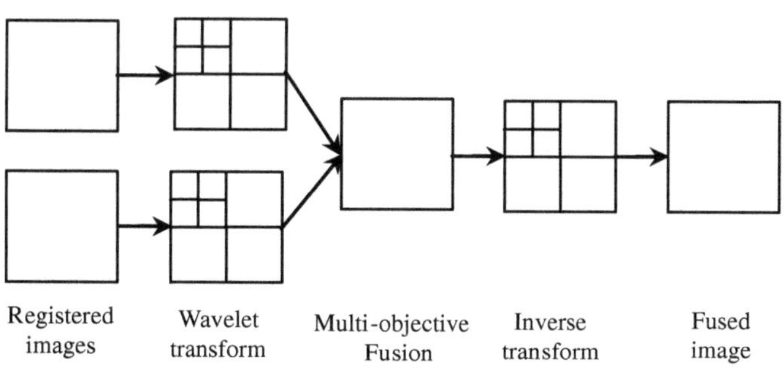

Fig. 7.3. Illustration of multi-objective pixel-level image fusion based on DWT

$$S_i(x,y) = \sum_m \sum_n W_j^2(x+m, y+n), j = 1, \ldots, J \tag{7.18}$$

where $W_j(x,y)$ is the wavelet coefficient at location (x,y), and (m,n) defines a window of coefficients around the current coefficient. The size of the window is typically small, e.g. 3 by 3. The coefficient with the largest salience is substituted for the fused coefficient while the less salient coefficient is discarded. The selection mode is implemented as

$$W_{Fj}(x,y) = \begin{cases} W_{Aj}(x,y), & S_{Aj}(x,y) \geq S_{Bj}(x,y), \\ W_{Bj}(x,y), & \text{otherwise.} \end{cases} \tag{7.19}$$

where $W_{Fj}(x,y)$ are the final fused coefficient in DWT domain, W_{Aj} and $W_{Bj}$are the current coefficients of A and B at level j .

Step 3: For approximations in DWT domain, use weighted factors to calculate the approximation of the fused image of F. Let C_F, C_A, and C_B be the approximations of F, A, and B respectively, two different fusion rules will be adopted. One rule called "uniform weight method (UWM)" is given by

$$C_F(x,y) = w_1 \cdot C_A(x,y) + w_2 \cdot C_B(x,y) \tag{7.20}$$

where the weighted factors of w_1 and w_2 are the values in the range of $[0,1]$, and they are also decision variables.

The other called "adaptive weight method (AWM)" is given by

$$C_F(x,y) = w_1(x,y) \cdot C_A(x,y) + w_2(x,y) \cdot C_B(x,y) \tag{7.21}$$

where $w_1(x,y)$ and $w_2(x,y)$ are decision variables.

Using a multi-objective optimization algorithm, we can find the optimal decision variables of multi-objective pixel-level image fusion in DWT domain, and realize the optimal image fusion.

Step 4: The new sets of coefficients are used to find the inverse transform to get the fused image F.

7.4 Evaluation Metrics of Image Fusion

In our method of image fusion, the establishment of an evaluation metric system is the basis of the optimization that determines the quality of the final fused image. However, in the image fusion literature only a few metrics for quantitative evaluation of different image fusion methods have been proposed. Generally, the construction of the perfect fused image is an illdefined problem since in most case the optimal combination is not known in advance.

In fact, the evaluation metrics of image fusion include subjective metrics and objective metrics. The subjective metrics rely on the ability of people's comprehension and are hard to come into application. While the objective metrics can overcome the influence of human vision, mentality and knowledge, and make machines automatically select a superior algorithm to accomplish the fusion mission.

We have explored the possibility to establish an objective evaluation metric system and got some meaningful results [9]. The objective metrics can be divided into three categories according to the reflected subjects. One category reflects the image features, such as gradient and entropy; the second reflects the relation of the fused image to the reference image, such as correlation coefficient, peak signal to noise ratio (PSNR); the third reflects the relation of the fused image to the source images, such as mutual information. As the optimization objectives, the greater the values of objective metrics are, the better the fused image is.

7.4.1 Image Feature Metrics

Gradient

Gradient reflects the change rate in image details that can be used to represent the clarity degree of an image. The higher the gradient of the fused image is, the clearer it is. Gradient is given by

$$G = \frac{\sum_{i=1}^{M-1} \sum_{j=1}^{N-1} \sqrt{[F(i,j) - F(i+1,j)]^2 + [F(i,j) - F(i,j+1)]^2}}{\sqrt{2}\,(M-1)(N-1)} \tag{7.22}$$

where M and N are the numbers of the row and column of image F respectively.

Entropy

Entropy is an metric to evaluate the information quantity contained in an image. If the value of entropy becomes higher after fusing, it indicates that the information quantity increases and the fusion performance is improved. Entropy is defined as

$$H = -\sum_{i=0}^{L-1} p_i \log_2 p_i \tag{7.23}$$

where L is the total of grey levels, p_i is the probability distribution of level i.

7.4.2 Image Similarity Metrics

Structural Similarity

Wang simulated the human vision system and proposed an metric of structural similarity which is better than correlation coefficient, for even if the fused image and the reference image are linearly related, there still might be relative distortions between them [21],[22]. Structural similarity (SSIM) is designed by modeling any image distortion as a combination of three factors: structure distortion, luminance distortion, and contrast distortion. SSIM is defined as

$$SSIM = \frac{\sigma_{FR} + C_1}{\sigma_F \sigma_R + C_1} \cdot \frac{2\mu_F \mu_R + C_2}{\mu_F^2 + \mu_R^2 + C_2} \cdot \frac{2\sigma_F \sigma_R + C_3}{\sigma_F^2 + \sigma_R^2 + C_3} \tag{7.24}$$

where μ_F and μ_R are the mean intensity of the fused image F and the reference image R respectively, σ_F and σ_RR are the standard deviation of F and R, σ_{FR} is the covariance, C_1, C_2, and C_3 are positive constant to avoid instability when denominators are very close to zero.

In (7.24), the first component is the correlation coefficient of F and R. The second component measures how close the mean gray levels of F and R is, while the third measures the similarity between the contrasts of F and R. The higher the value of SSIM is, the more similar to R the F is. The dynamic range of SSIM is [-1, 1]. If two images are identical, the similarity is maximal and equals 1; while one is the negative of the other, SSIM equals -1.

Peak Signal to Noise Ratio

The higher the value of PSNR is, and the lower the value of RMSE is, the better the fused image is. PSNR is defined as

$$PSNR = 10 \lg \frac{255^2}{RMSE^2} \tag{7.25}$$

where RMSE (root mean squared error) is defined as

$$RMSE^2 = \frac{1}{MN} \sum_i \sum_j [R(i,j) - F(i,j)]^2 \tag{7.26}$$

7.4.3 Mutual Information Metrics

Conditional Mutual Information

Qu[23], Ramesh[24] and Wang[25] et al. adopted mutual information to represent the amount of information that is transferred from the source images to the final fused image, where no attention has been paid to the overlapping information of the source images, so this metric can't effectively evaluate the mutual information among the fused image and the source images. Vassilis[26] proposed the conditional mutual information which can avoid overlapping information of the source images. However, the calculation expression is a bit complex. We make an improvement and get new conditional mutual information.

The relationship among entropy and mutual information for three variables of X_1, X_2, and F is demonstrated by the Venn diagram of Fig. 7.4.

According to Fig. 7.4, the overlapping information of I_0 transferred from the source images to the final fused image in [23] is given by

$$\begin{aligned} I_0 &= I(X_1; F) - I(X_1; F|X_2) \\ &= I(X_2; F) - I(X_2; F|X_1) \end{aligned} \tag{7.27}$$

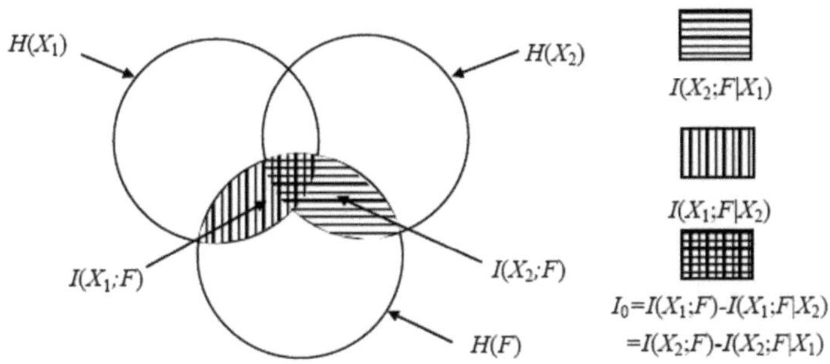

Fig. 7.4. Relationship among entropy and mutual information for three variables

The conditional mutual information of random variables X_2 and F given X_1 is defined by [26]

$$\begin{aligned} I(X_2; F|X_1) &= H(X_2|X_1) - H(X_2|F, X_1) \\ &= H(X_1; X_2) - H(X_1, X_2|F) - I(|X_1; F) \end{aligned} \tag{7.28}$$

where $H(X_2|X_1)$ denotes the conditional entropy of X_2 given X_1, $I(X_1; F)$ denotes the mutual information of X_1 and F.

The sum of all the conditional information (CI) transferred from the source images X_i to the final fused image F and is expressed as

$$\begin{aligned} CI &= I(X_1, \ldots, X_n; F) \\ &= I(X_1, F) + I(X_2; F|X_1) + \ldots + I(X_n; F|X_{n-1}, \ldots, X_1) \end{aligned} \tag{7.29}$$

According to (7.28) and (7.29), we can get the following simplified expression of CI

$$CI = H(X_1, \ldots, X_n) - H(X_1, \ldots, X_n|F) \tag{7.30}$$

In fact, the difference of the federal entropy in the source images and the condition entropy given the fused image is the effective mutual information, which can avoid overlapping information.

The overlapping information of I_0 for two source images can be obtained by

$$I_0 = I(X_1; F) + I(X_2; |F) - CI \tag{7.31}$$

In order to make CI be bounded, we define the unitary conditional mutual information (CMI) as

$$CMI = \frac{CI}{H(X_1, \ldots, X_n)} = 1 - \frac{H(X_1, \ldots, X_n|F)}{H(X_1, \ldots, X_n)} \tag{7.32}$$

where the denominator of $H(X_1, \ldots, X_n)$ denotes the federal entropy of $X_1, \ldots, X_n$ and defined as

$$H(X_1, \ldots, X_n) = - \sum_{x_1, \ldots, x_n} p(x_1, \ldots, x_n) \log p(x_1, \ldots, x_n) \tag{7.33}$$

The numerator of $H(X_1,\ldots,X_n|F)$denotes the conditional entropy of $X_1,\ldots,X_n$ given F and defined as

$$H(X_1,\ldots,X_n|F) = -\sum_{x_1,\ldots,x_n,f} p(x_1,\ldots,x_n,f)\log\frac{p(x_1,\ldots,x_n)}{p(f)} \tag{7.34}$$

A higher value of CMI indicates that the fused image contains fairly good quantity of information presented in both the source images. CMI takes values in the range [0, 1], where zero corresponds to total lack of common information between the source images and the fused image and one corresponds to an effective fusion process that transfers all the information from the source images to the fused image (in the ideal case).

Information Symmetry

A high value of CMI doesn't imply that the information from both the images is symmetrically fused, e.g. when F is identical with X_1, CI will be high and take the value of $H(X_1)$. Therefore, we introduce the metric of information symmetry (InfS) from [24] and make an improvement. We should use the difference between $I(X_2;F|X_1)$ and $I(X_1;F|X_2)$to measure the degree of information symmetry. Since the following expression is valid

$$I(X_2;F|X_1) - I(|X_1;F|X_2) = I(X_2;F) - I(X_1;F) \tag{7.35}$$

In order to simplify the expression, define InfS as

$$InfS = 1 - \left|\frac{I(X_2;F) - I(X_1,F)}{max[I(X_1;F), I(X_2,F)]}\right| \tag{7.36}$$

InfS is an indication of how much symmetric the fused image is, with respect to input images. The higher the value of InfS is, the better the fusion result is. InfS also takes values in the range [0, 1], where zero implies that the fused image is identical with some one of source images, while one implies that both the images is symmetrically fused.

7.5 MOCPSO Algorithm

J. Kennedy and R. C. Eberhart brought forward particle swarm optimization (PSO) inspired by the choreography of a bird flock in 1995[27]. Unlike conventional evolutionary algorithms, PSO possesses the following characteristics: 1) Each individual (or particle) is given a random speed and flows in the decision space; 2) each individual has its own memory; 3) the evolution of each individual is composed of the cooperation and competition among these particles. Since the PSO was proposed, it has been of great concern and becomes a new research field[28]. PSO has shown a high convergence speed in single objective optimization, and it is also particularly suitable for multi-objective optimization [14],[29],[30]. In order to improve the performances of the algorithm, we present a proposal, called "multi-objective constriction particle swarm optimization" (MOCPSO) using MOPSO and NSGA-II for reference, in which a new crowding operator is used to improve the distribution of nondominated solutions along the Pareto front and maintain the population diversity; an adaptive mutation operator are introduced to improve the search capacities and avoid the earlier convergence; the uniform design is used to obtain the optimal combination of the algorithm parameters.

7.5.1 MOCPSO Flow

The algorithm of MOCPSO is shown in Fig. 7.5. First the position and velocity of each particle in the population are initialized, and the nondominated particles are stored in the repository. Second the velocity and position of each particle are updated, the partly particles mutate and the particles are maintained within the decision space. When a decision variable goes beyond its boundaries, the decision variable takes the value of its corresponding boundary, then its velocity is multiplied by (-1) so that it searches in the opposite direction. Third each particle is evaluated and their records and the repository are updated; then the cycle begins. When the maximum cycle number is reached, the Pareto solutions in the repository are output.

7.5.2 Initialization of Algorithm

Step 1. Initialize the position of each particle with arbitrary where the particles called decision variables denotes the fusion parameters;

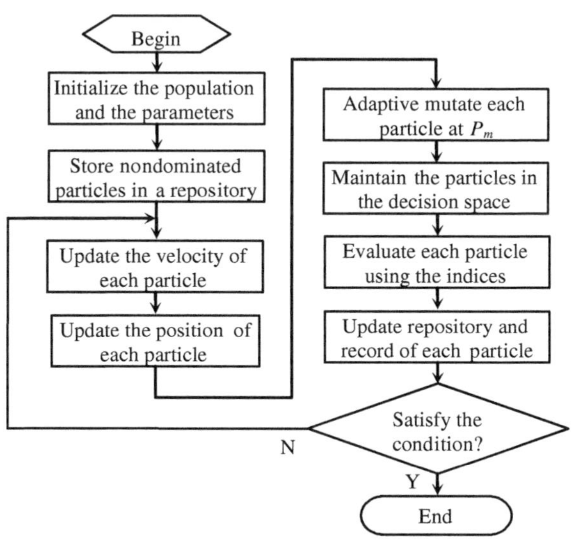

Fig. 7.5. Illustration of MOCPSO algorithm

$$pop[i] = \text{arbitrary value}, i = 1, \ldots, N_p \tag{7.37}$$

where N_p is the number of particles.
Step 2. Initialize the velocity of each particle with zero where the velocities denote the changes of the parameters;

$$vel[i] = 0 \tag{7.38}$$

Step 3. Initialize the record of each particle with the current position where the record denotes the searched best particle;

$$Pbest[i] = pop[i] \tag{7.39}$$

where *Pbest* is the searched best position for the ith particle.
Step 4. Evaluate each of the particles where the fitness called optimization objectives are the evaluation metrics of image fusion, get

$$fun[i, j], i = 1, \ldots, N_p, j = 1, \ldots, N_f \tag{7.40}$$

where N_f is the number of objectives.
Step 5. Store the nondominated particles in the external Repository of REP according to the Pareto optimality

$$REP = \{rep[i], \ldots, rep[N_M]\} \tag{7.41}$$

where M is the allowed maximum capacity of REP.

7.5.3 Update Particle Swarm

Update the velocity and the position of each particle using following canonical form.

$$\begin{cases} vel[i+1] = \chi \cdot (vel[i] + \varphi \cdot (pop_m[i] - pop[i])) \\ pop[i+1] = pop[i] + vel[i+1] \\ pop_m[i] = (\varphi_1 \cdot Pbest[i] + \varphi_2 \cdot rep[h]) / (\varphi_1 + \varphi_2) \end{cases} \tag{7.42}$$

where $\chi$is constriction factor that causes convergence of the individual trajectory in the search space and whose value is typically approximately 0.7289 [31], $\varphi = (\varphi_1 + \varphi_2)$; φ_1 and φ_2 are the acceleration coefficients with random values in the range [0, 2.05]; $Pbest[i]$ is the best position that the particle i has had; h is the index of the solution in the repository with maximum crowding distance that implies the particle locates in the sparse region, as aims to maintain the population diversity; $pop[i]$ is the current position of particle i. $vel[i]$ is the current velocity of the particle i.

7.5.4 Adaptive Mutation

PSO is known to have a very high convergence speed. However, such a convergence speed may be harmful in the context of multi-objective optimization. An adaptive mutation operator is applied not only to the particles of the swarm, but also to the range of each design variable of the problem[14]. What this does is to cover the full range of each design variable at the beginning of the search and then we narrow the range covered over time, using a nonlinear function.

$$R = (Upper - Lower) \cdot (1 - \frac{g}{p_m G_{max}})^2 \tag{7.43}$$

where Upper is the upper of the design variable, Lower is the lower, R is the value range of the variables.

7.5.5 Repository Control

Control Strategy

The external repository of REP is used to record the nondominated particles in the primary population of our algorithm. At the beginning of the search, the REP is empty. The nondominated particles found at each iteration are compared with respect to the contents of the REP. If the REP is empty, the current solution will be accepted. If this new solution is dominated by an individual within the REP, such a solution will be automatically discarded. Otherwise, if none of the elements contained in the REP dominates the solution wishing to enter, such a solution will be stored in the REP. If there are solutions in the REP that are dominated by a new element, such solutions will be removed out of the REP. Finally, if the REP has reached its allowed maximum capacity, the new nondominated solution and the contents of the REP will be combined into a new population, according to the objectives, the individuals with lower crowding distances (locating the dense region) will not enter into the REP.

Crowding Operator

In order to improve the distribution of nondominated solutions along the Pareto front, we introduce a concept of crowding distance from NSGA-II[12] that indicates the population density. When comparing the Pareto optimality between two individuals, we find that the one with a higher crowding distance (locating the sparse region) is superior. In [12], the crowding distance is defined as the size of the largest cuboids enclosing the point i without including any other point in the population, and it can be acquired through calculating average distance of two points on either side of point of the objective. However, the definition has $O(mn \log n)(m = N_f, n = N_p)$ computational complexity, and may need too much time because of sorting order.

Here we propose a new crowding distance that can be calculated using level sorting, and doesn't need to sort order for each objective. The crowding distance of the boundary points is set to infinity so that they can be selected into the next generation. The others can be calculated with respect to their objectives. For objective j, we divide its value range into a special level according to the boundary, then sort these levels in a descending order of the particle numbers, and compute the crowding distance using

$$d_{ij} = \frac{S_{ij}}{N_{ij}} \tag{7.44}$$

where d_{ij} is the crowding distance of particle i at objective j, S_{ij} is the sequence number of the level where particle i locates, N_{ij} is the number of the particles in level S_{ij} . The crowding distance of particle i is defined as

$$Dis[i] = \sum_j d_{ij} \tag{7.45}$$

The level number should be chosen carefully. If it is too small, many particles will have identical crowding distances. If it is too large, the computation time will be comparable to that of the original crowding distance computation algorithm in NSGA-II. It should be also stated that the crowding distances estimated by the proposed algorithm is worse than those calculated by the original algorithm.

The new crowding distance doesn't need to sort order for every objective and has less complexity, and it is superior to the grid [11], [14] because the latter may fail to allocate memory when there exist too many objectives.

7.5.6 Uniform Design for Parameter Establishment

As in the case of many evolutionary optimization algorithms, the studies have concluded that the performance of the PSO is sensitive to control parameter choices. So the algorithm parameters must be established correctly. The relatively important parameters of MOCPSO include the number of particles, the number of cycles, the size of the repository, the constriction factor, and the mutation probability. In order to attain the optimal combination of these parameters, we introduce the uniform design.

Uniform design is used to convert the problem of parameter establishment into the experimental design of multi-factor and multi-level, which can reduce the work load of experiment greatly of simulation[32]. The main objective of uniform design is to sample a small set of points from a given set of points, such that the sampled points are uniformly scattered. Let there be n factors and $q$levels per factor. When n and q are given, the uniform design selects q combinations out of q^n possible combinations, such that these q combinations are scattered uniformly over the space of all possible combinations. The selected combinations are expressed in terms of a uniform array $U(n,q) = [U_{i,j}]_{q \times n}$, where $U_{i,j}$ is the level of the jth factor in the ith combination. When q is prime and $q > n$, $U_{i,j}$ is given by

$$U_{i,j} = (i\sigma^{j-1} \text{mod} q) + 1 \tag{7.46}$$

where σ is a parameter determined by the number of factors and the number of levels per factor, and the value of σ is given in Table 7.1[33].

Table 7.1. σ values for different number of factors and different number of levels

Level Number	Factor Number	σ
5	2-4	2
7	2-6	3
11	2-10	7
13	2	5
13	3	4
13	4-12	6
17	2-16	10
19	2-3	8
19	4-18	14
23	2, 13-14, 20-22	7
23	8-12	15
23	3-7, 15-19	17
29	2	12
29	3	9
29	4-7	16
29	8-12, 16-24	8
29	13-15	14
29	25-28	18
31	2, 5-12, 20-30	12
31	3-4, 13-19	22

7.5.7 Convergence Analysis of MOCPSO

Recently, some theoretical studies about the convergence properties of PSO have been published[34]. Most of the theoretical studies are based on simplified PSO models, in which a swarm consisting of one particle of one dimension is studied. The *Pbest* and *Gbest* particles are assumed to be constant throughout the process. Convergence about PSO has been defined as follows:

Definition 1. *Considering the sequence of global best solutions* $\{Gbest_t\}_{t=0}^{\infty}$, *we say that the swarm converges iff*

$$\lim_{t\to\infty} Gbest_t = p \tag{7.47}$$

where p *is an arbitrary position in the search space.*

Since p refers to an arbitrary solution, Definition 1 does not mean convergence to a local or global optimum. Van den Bergh [35] concluded (assuming uniform distributions) that the particle then converges to the position:

$$(1-a)Pbest + aGbest \tag{7.48}$$

where $a = c_2/(c_1 + c_2)$, c_1 and c_2 are learning factors (accelerating coefficients).

It can be seen that a particle converges to a weighted average between its personal best and its neighborhood best position. However, we can only ensure the convergence of PSO to the best position visited by all the particles of the swarm. In order to ensure convergence to the local or global optimum, two conditions are necessary:

(1) The $Gbest_{t+1}$ solution can be no worse than the $Gbest_t$ solution (monotonic condition);
(2) The algorithm must be able to generate a solution in the neighborhood of the optimum with nonzero probability, from any solution x of the search space.

To the best of our knowledge, there are few studies about the convergence properties of MOPSOs. From the discussion previously provided, it can be concluded that it is possible to ensure convergence, by correctly setting the parameters of the flight formula. But, as in the case of single-optimization, such property does not ensure the convergence to the true Pareto front in the case of multi-objective optimization. In this case, condition (1) must change to [34]:

(1) The solutions contained in the external archive at iteration $t+1$ should be nondominated with respect to the solutions generated in all iterations τ, $0 \leq \tau \leq t+1$, so far (monotonic condition).

But the normal dominance-based strategies do not ensures this condition, unless they make sure that for any solution discarded from the repository one with equal or dominating objective vector is accepted. In this way, the proposed MOCPSO approach satisfies condition (1) according the control strategy, and also satisfies condition (2) according the algorithm designs. Thus, MOCPSO can remain to explore and ensure global convergence to the true Pareto front in sufficient time.

7.6 Experiments and Analysis

The experiments include algorithm test, multi-focus image fusion, blind image fusion, multi-resolution image fusion and color image fusion. In this section, the optimization algorithms are compared firstly, then the performances of the proposed multi-objective pixel-level image fusion method are tested and compared with that of different fusion schemes. We use MOCPSO to search the Pareto optimal weights of the image fusion model and compare the results with those of simple wavelet method (SWM) that takes a fixed fusion weight of 0.5 for the approximations. The sum of the weights at each position of two source images is limited to 1. All approaches are run for a maximum of 100 evaluations, and the results in the tables are the average of these evaluations. In the context, AWM denotes the AWM based on MOCPSO, UWM denotes UWM based on MOCPSO.

Since the solutions to multi-objective pixel-level image fusion are nondominated by one another, we give the preference for the evaluation metrics so as to select the Pareto optimal solutions to compare. When the reference image exists, the order of preference is SSIM, CMI, Entropy, PSNR, Gradient, and InfS. SSIM is the principal objective, the higher the value of SSIM is, and the more similar the fused image to the reference image, and the better the results are. If the reference image doesn't exist, the order is CMI, Entropy, Gradient and InfS.

7.6.1 Uniform Design for MOCPSO Parameters

In order to make MOCPSO perform well, the uniform design is firstly used to attain the optimal combination of algorithm parameters. We construct a uniform array with five factors and seven levels as follows, where σ is equal to 3. We compute $U(5,7)$ based on (7.46) and get

$$U(5,7) = \begin{bmatrix} 2 & 4 & 3 & 7 & 5 \\ 3 & 7 & 5 & 6 & 2 \\ 4 & 3 & 7 & 5 & 6 \\ 5 & 6 & 2 & 4 & 3 \\ 6 & 2 & 4 & 3 & 7 \\ 7 & 5 & 6 & 2 & 4 \\ 1 & 1 & 1 & 1 & 1 \end{bmatrix} \tag{7.49}$$

In the first combination of (7.49), the four factors have respective levels 2, 4, 3, 7, 5; in the second combination, the four factors have respective levels 3, 7, 5, 6, 2, etc. The value range of the number of particles is [20, 200]; the range of the number of cycles is [50, 350]; the range of the size of the repository is [70, 250]; the range of the constriction factor is [0.70, 0.76]; the range of the mutation probability is [0.01, 0.07].

We introduce some evaluation criteria, including Objective Distance(OD) [36], Inverse Objective Distance(IOD) [37] , Spacing (SP)[38] and Error Ratio (ER) [39] to evaluate the performance of MOCPSO. OD is the average distance from each solution in a nondominated solution set to its nearest Pareto-optimal solution, IOD is the average distance from each Pareto-optimal solution to its nearest solution in the nondominated solution set, SP is the distance variance of neighboring solutions in the nondominated solution set found so far, ER is the percentage of solutions from the nondominated solution set found so far that are not members of the true Pareto optimal set.

We use the following test function [40] to attain the optimal parameter combination.

$$\begin{aligned} &\max f_1(\overrightarrow{x}) = -x_1^2 + x_2 \\ &\max f_2(\overrightarrow{x}) = x_1/2 + x_2 + 1 \\ &s.t. \;\; x_1/6 + x_2 - 13/2 \leq 0 \\ &\qquad x_1/2 + x_2 - 15/2 \leq 0 \\ &\qquad 5/x_1 + x_2 - 30 \leq 0 \\ &\qquad 0 \leq x_1, x_2 \leq 7 \end{aligned} \tag{7.50}$$

Table 7.2. Evaluation criteria of different combinations

Com	OD	IOD	SP	ER
U1	0.00422	0.00406	0.07634	0.2165
U2	0.00965	0.00567	0.15488	0.3461
U3	0.01773	0.01454	0.35632	0.3684
U4	0.00302	0.00399	0.08363	0.2316
U5	0.00148	0.00156	0.03542	0.1043
U6	0.00587	0.00998	0.28643	0.2457
U7	0.64291	0.06025	1.47912	0.5426

All combinations are run for a maximum value of 100 evaluations. As shown in Table 7.2, results indicate that the fifth combination is the optimal in the problem.

By the uniform design, the parameters of MOCPSO are as follow: the particle number of N_p is 170; the maximum cycle number of G_{max} is 100; the allowed maximum capacity of MEM is 160; the value of the constriction factor is 0.72; the mutation probability of P_m is 0.07. The inertia weight of W_{max} is 1.2, and W_{min} is 0.2; the learning factor of c_1 is 1, and c_2 is 1, the parameters of MOPSO are the same, while the inertia weight of W is 0.4, the grid number of N_{div} is 20, for a greater number may cause the failure of program execution, e.g. 30.

7.6.2 Comparison of MOCPSO and MOPSO

The performances of MOCPSO are tested using the experimental methods. The results of MOCPSO is compared with NSGA-II, MOPSO et al. Table 7.3, Table 7.4, Table 7.5, Table 7.6, and Table 7.7 show the comparison of results among different algorithms, including NSGA-II, MOPSO, where MOCPSO I denotes the MOCPSO with a linear inertia weight[41] but not with the constriction form, MOCPSO II denotes the MOCPSO without the mutation operator, MOCPSO III denotes the MOCPSO with the crowding operator of NSGA-II.

It can be seen that the metrics of MOCPSO I, MOCPSO II, and MOCPSO III are inferior to those of MOCPSO, which indicates that the constriction form is better than the inertia weight, the mutation operator can avoid earlier convergence and improve the search capacities; the new crowding operator can increase the running speed and improve the distribution of nondominated solutions along the Pareto front. The metrics of MOPSO and NSGA-II are inferior to those of MOCPSO, which indicates that MOCPSO has better search capacity and faster convergent speed than MOPSO, where MOPSO needs too much memory and time, for the grid is worse for too many objectives. The results show that MOCPSO can effectively avoid earlier convergence and improve the search capacities, especially while there are too many objectives and too many variables.

7.6.3 Multi-focus Image Fusion

Multi-focus image fusion is an important research area in pixel-level image fusion which can attain an all in-focus merged image from multiple images with different focuses and the same scene. The image "plane" from CCITT is selected as the reference image of R with the 256 × 256 pixels in size. Through image processing, we get two source images of A and B. As shown

Table 7.3. Results of Objective Distance of different algorithms

OD	NSGA-II	MOPSO	MOCPSOI	MOCPSOII	MOCPSOIII	MOCPSO
Best	0.003885	0.002425	0.001638	0.002693	0.001241	0.000815
Worst	0.678449	0.476815	0.292582	0.317468	0.242675	0.209670
Average	0.084239	0.036535	0.011537	0.014325	0.005264	0.001689
Median	0.011187	0.007853	0.006415	0.007139	0.003347	0.001527
Std. Dev.	0.165244	0.104589	0.071724	0.086952	0.062513	0.048752

Table 7.4. Results of Inverse Objective Distance of different algorithms

IOD	NSGA-II	MOPSO	MOCPSOI	MOCPSOII	MOCPSOIII	MOCPSO
Best	0.004279	0.003654	0.001815	0.002784	0.001463	0.000975
Worst	0.712548	0.427437	0.308864	0.328573	0.267982	0.221684
Average	0.092386	0.032842	0.012212	0.015889	0.005366	0.002134
Median	0.014860	0.008326	0.009233	0.007432	0.003912	0.001950
Std. Dev.	0.186383	0.090149	0.077342	0.091891	0.065238	0.052718

Table 7.5. Results of Spacing of different algorithms

SP	NSGA-II	MOPSO	MOCPSOI	MOCPSOII	MOCPSOIII	MOCPSO
Best	0.001032	0.043982	0.008742	0.016348	0.002039	0.002897
Worst	1.488681	0.538102	0.468721	0.426638	0.366758	0.297314
Average	0.098486	0.109452	0.076942	0.085891	0.069971	0.046496
Median	0.027173	0.067481	0.053478	0.061452	0.041245	0.021548
Std. Dev.	0.327387	0.110051	0.097232	0.106847	0.092693	0.071471

Table 7.6. Results of Error Ratio of different algorithms

ER	NSGA-II	MOPSO	MOCPSOI	MOCPSOII	MOCPSOIII	MOCPSO
Best	0.7532	0.0813	0.0742	0.0663	0.0476	0.0368
Worst	0.9924	0.2721	0.2493	0.2975	0.2347	0.2429
Average	0.8965	0.1325	0.1346	0.1484	0.1296	0.1045
Median	0.9216	0.1432	0.1468	0.1505	0.1482	0.1381
Std. Dev.	0.0671	0.0450	0.0271	0.0389	0.0205	0.0193

Table 7.7. Computational time (in seconds) of different algorithms

time	NSGA-II	MOPSO	MOCPSOI	MOCPSOII	MOCPSOIII	MOCPSO
Best	0.9892	0.0071	0.0063	0.0078	0.0052	0.0045
Worst	1.1352	0.2722	0.2685	0.2862	0.2584	0.2396
Average	1.0882	0.2513	0.2136	0.2272	0.2362	0.1102
Median	1.1085	0.2642	0.2381	0.2407	0.2413	0.1358
Std. Dev.	0.0432	0.0575	0.0362	0.0394	0.0305	0.0326

(a) Image A (b) Image B (c) UMW image (d) AWM image

Fig. 7.6. Results of multi-objective multi-focus image fusion

in Fig. 6(a) and Fig. 6(b), the left region of A is blurred, while the right region of B is blurred. The fused images from the Pareto optimal solutions are shown in Fig. 6(c) and Fig. 6(d) using UWM and AWM at level 3 in DWT domain.

Table 7.8 shows the evaluation metrics of the fused images from different schemes. From Table 7.8, we can see that when the decomposition level equals zero in DWT domain, which implies that the fusion is performed in the spatial domain, the metrics of AWM are inferior to those of UWM. The reason is that the decision variables of AWM in spatial domain are too many and AWM can't reach the Pareto optimal front in a limited time, e.g. the number of iteration is 100. The run time of AWM must increase with the number of decision variables, so AWM can only be regarded as an ideal method of image fusion in spatial domain. The advantage of spatial fusion is easy to realize, however the simple splice of pixels smooths the image and is not convenient for the later processing, such as image comprehension.

In DWT domain, the metrics of AWM at level 3 are superior to those of AWM at other levels. The higher the decomposition level is, the better the fused image is, for a higher level decreases the decision variables and improves the adaptability. Moreover, the metrics of AWM are superior to those of UWM because the weights of AWM are adaptive in different regions. The metrics of SWM are inferior to our results except InfS. InfS can't be used as an important objective, for InfS may have reached the maximum before fusing.

Table 7.8. Evaluation metrics of multi-objective multi-focus image fusion

Schemes	Level	Entropy	Gradient	CMI	InfS	PSNR	SSIM
UWM	0	6.6826	6.3220	0.5448	0.9995	29.1992	0.9794
AWM	0	6.6929	6.7477	0.5452	0.9995	28.8614	0.9778
AWM	1	6.6433	8.2222	0.5579	0.9992	29.7771	0.9824
AWM	2	6.6591	8.9455	0.5587	0.9993	31.9960	0.9900
UWM	3	6.6675	9.1465	0.5620	0.9993	34.6555	0.9952
SWM	3	6.6557	8.3813	0.5404	0.9994	34.4461	0.9859
AWM	3	6.6849	9.1868	0.5693	0.9993	35.6626	0.9965

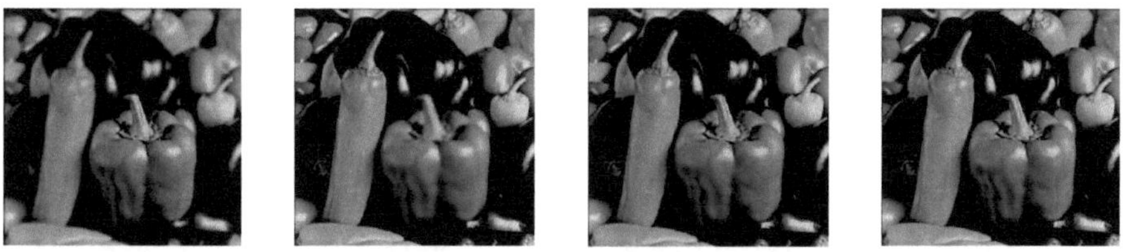

(a) Image A (b) Image B (c) UWM image (d) AWM image

Fig. 7.7. Results of multi-objective blind image fusion

7.6.4 Blind Image Fusion

Blind image fusion denotes the category of image fusion without the reference image, which is very popular in practice. In the situation, the method of multi-objective pixel-level image fusion needs to optimize the fusion parameters according to the metrics of CMI, entropy, gradient, and InfS. The image "pepper" with the 256×256 pixels in size is selected to test the performance of the method. The source images of A and B are shown in Fig. 7(a) and Fig. 7(b), where the background of A is fuzzy, and the foreground of B is fuzzy. The optimal fused image from the Pareto optimal solutions are shown in Fig. 7(c) and Fig. 7(d) using UWM and AWM at level 3 in DWT domain.

Table 7.9 shows the evaluation metrics of the fused images from different schemes. It can be seen that the method of multi-objective blind image fusion show its effectiveness. Under the direction of conditional mutual information, this method acquires most information from the source images, symmetrically fuses the source images, and increases the information and the definition in the fused image. Table 7.9 also indicates that MOCPSO performs well.

7.6.5 Multi-resolution Image Fusion

It often arises that the source images have different resolutions. The multi-resolution image fusion techniques merge the spatial information from a high-resolution image with the radiometric information from a low-resolution image to improve the quality of the fused image. The image "aerial" is selected

Table 7.9. Evaluation metrics of multi-objective blind image fusion

Schemes	Level	Entropy	Gradient	CMI	InfS
UWM	0	7.5296	6.2450	0.5362	0.9992
AWM	0	7.5269	6.5097	0.5680	0.9994
AWM	1	7.5326	7.1619	0.5683	0.9992
AWM	2	7.5345	7.1627	0.5697	0.9992
UWM	3	7.5328	7.1679	0.5698	0.9991
SWM	3	7.5216	7.1223	0.5616	0.9990
AWM	3	7.5424	7.2153	0.5765	0.9995

(a) Image A (b) Image B (c) UMW image (d) AWM image

Fig. 7.8. Results of multi-objective multi-resolution image fusion

as the reference image of R. The two source images of A and B are shown in Fig. 8(a) and Fig. 8(b) respectively, where A is the low-resolution image (128×128) with the pixel size of 30 m, but clear, while B is the high-resolution image (256×256) with the pixel size of 15 m, but blurred.

The decomposition level of the high-resolution image is smaller than that of the low-resolution image in DWT domain. When the decomposition level of the former equals 1, the level of the latter will equal 0, and the coefficients of the high-resolution image will be substituted for the fused coefficient. In other levels, the method is the same as the text. The optimal fused image from the Pareto optimal solutions are shown in Fig. 8(c) and Fig. 8(d) using

Table 7.10. Evaluation metrics of multi-objective multi-resolution image fusion

Schemes	(J_A, J_B)	Entropy	Gradient	CMI	INFS	PSNR	SSIM
UWM	(0, 1)	7.1529	12.1093	0.3144	0.9825	27.5151	0.9115
AWM	(0, 1)	6.9679	14.2213	0.3392	0.9841	15.8836	0.8076
UWM	(1, 2)	6.8752	11.6481	0.3330	0.9807	24.1379	0.9280
AWM	(1, 2)	7.1623	12.1546	0.3480	0.9825	27.5682	0.9621
UWM	(2, 3)	6.9595	11.9852	0.3516	0.9819	24.9924	0.9414
SWM	(2, 3)	7.1340	11.9730	0.3435	0.9806	25.2568	0.9205
AWM	(2, 3)	7.1823	12.4250	0.3556	0.9823	27.7075	0.9636

UWM and AWM at level 3 in DWT domain. Table 7.10 shows the evaluation metrics of the fused images from different schemes, where (J_A, J_B) denotes that the decomposition level of A is J_A, and that the decomposition level of B is J_B. It can be seen that the method of multi-objective multi-resolution image fusion using adaptive fusion weights based on DWT at level 3 of image B can get a clear fused image with a high resolution, and that MOCPSO also performs well.

7.6.6 Color Image Fusion

The color image fusion techniques fuse the color images to attain more color information and enhance the features of the image. The image "fishman" from Lab. for Image and Video Engineering of Texas University is selected as the reference image of R with 256×256 pixels in size, each pixel being represented by three bytes (one for each of the R, G, and B channels). The two source images of A and B are shown in Fig. 9(a) and Fig. 9(b). In order to fuse the color source images, we choose the YUV color space, which has components representing luminance, saturation, and hue, for the color components are dependent on each other in RGB color space. The conversion from RGB to YUV is given by

$$\begin{bmatrix} Y \\ U \\ V \end{bmatrix} = \begin{bmatrix} 0.299 & 0.587 & 0.114 \\ -0.148 & -0.289 & 0.437 \\ 0.615 & -0.515 & -0.100 \end{bmatrix} \begin{bmatrix} R \\ G \\ B \end{bmatrix} \tag{7.51}$$

Component Y represents the luminance, so the fusion is performed and the results are evaluated in the Y component. Since the source images can be assumed to have similar saturation and hue, the average of the U and V components from source images can be substituted for the U and V components in the fused image respectively. The optimal fused image from the Pareto optimal solutions are shown in Fig. 9(c) and Fig. 9(d) using UWM and AWM at level 3 in DWT domain. Table 7.11 shows the evaluation metrics of the fused images from different schemes. It can be seen that the method of multi-objective color image fusion based on DWT yields a clear color image and that MOCPSO also performs well.

(a) Image A (b) Image B (c) UMW image (d) AWM image

Fig. 7.9. Results of multi-objective color image fusion

Table 7.11. Evaluation metrics of multi-objective color image fusion

Schemes	Level	Entropy	Gradient	CMI	InfS	PSNR	SSIM
UWM	0	7.6890	8.8186	0.5226	0.9991	29.2640	0.9930
AWM	0	7.7108	9.3027	0.5283	0.9993	28.3477	0.9912
AWM	1	7.7271	11.8705	0.5357	0.9992	32.5909	0.9968
AWM	2	7.7237	11.8861	0.5362	0.9994	32.6471	0.9969
UWM	3	7.7286	12.0236	0.5383	0.9993	33.6572	0.9973
SWM	3	7.7214	11.9203	0.5327	0.9992	32.3347	0.9963
AWM	3	7.7394	12.0312	0.5396	0.9994	33.8427	0.9977

Therefore, the method of multi-objective pixel-level image fusion that uses MOCPSO to search the adaptive fusion weights at level 3 in DWT domain is the optimal. This method could save up the optic features of the images in contrast to the spatial method, and overcome the limitations of given fusion parameters.

7.7 Conclusions

Different from the conventional image fusion methods that the fusion is done before the evaluation, the proposed method of multi-objective pixel-level image fusion in this chapter with a view of the fusion objectives, design the proper evaluation metrics, then optimize the parameters using the multi-objective optimization algorithm and attain the optimal fusion results. Thus, the pixel-level image fusion method is simplified, and the limitations of much too dependence on the experience are overcome. Through analyzing the metric of mutual information, we define new metrics of conditional mutual information and information symmetry, and get a reasonably sound evaluation metric system. The proposed multi-objective constriction particle swarm optimization (MOCPSO) is an effective algorithm to solve the multi-objective problem, especially when the number of objectives is large, which can get to the Pareto front of optimization problems quickly and attain the optimal solutions. Experiments show that it is feasible and relatively effective to use MOCPSO to optimize the parameters of multi-objective pixel-level image fusion. MOCPSO can be effectively applied to solve other multi-objective problems, where the parameters of algorithm are the key for different optimization problem, and the effectiveness of algorithm can be improved further.

Acknowledgments. The authors wish to thank Prof. Carlos A. Coello Coello in CINVESTAV-IPN and Dr. Vassilis Tsagaris in University of Patras for their generous helps. We also thank Mr. Liang Xiaowei for his corrections to the chapter.

References

1. Pohl, C., Genderen, J.L.V.: Multisensor image fusion in remote sensing: concepts, methods and applications. Int. J. Remote Sens. 19(5), 823–854 (1998)
2. Piella, G.: A general framework for multiresolution image fusion: from pixels to regions. Inf. Fusion. 4(4), 259–280 (2003)
3. Petrovic, V.S., Xydeas, C.S.: Gradient-based multiresolution image fusion. IEEE Trans. Image Process 13(2), 228–237 (2004)
4. Choi, M., Kim, R.Y., Nam, M.R., et al.: Fusion of multispectral and panchromatic satellite images using the curvelet transform. IEEE Trans. Geosci. Remote Sens. Lett. 2(2), 136–140 (2005)
5. De, I., Chanda, B.: A simple and efficient algorithm for multifocus image fusion using morphological wavelets. Signal Process 86(5), 924–936 (2006)
6. Wang, Z.J., Ziou, D., Armenakis, C., et al.: A comparative analysis of image fusion methods. IEEE Trans. Geosci. Remote Sens. 43(6), 1391–1402 (2005)
7. Nasrabadi, N.M., Clifford, S., Liu, Y.: Integration of stereo vision and optical flow using an energy minimization approach. J. Opt. Sot. Amer. A 6(6), 900–907 (1989)
8. Qin, Z., Bao, F.M., Li, A.G., et al.: 2004 Digital image fusion. Xi'an Jiaotong University Press, Xi'an (in Chinese)
9. Niu, Y.F., Shen, L.C.: A novel approach to image fusion based on multi-objective optimization. In: Proceedings of IEEE WCICA 2006, Dalian, pp. 9911–9915 (2006)
10. Knowles, J.D., Corne, D.W.: Approximating the nondominated front using the pareto archived evolution strategy. Evol. Comput. 8(2), 149–172 (2000)
11. Zitzler, E., Laumanns, M., Thiele, L.: SPEA2: improving the strength pareto evolutionary algorithm. Technical Report 103, Gloriastrasse 35, CH-8092 Zurich, Switzerland (2001)
12. Deb, K., Pratap, A., Agarwal, S., et al.: A fast and elitist multiobjective genetic algorithm: NSGA-II. IEEE Trans. Evol. Comput. 6(2), 182–197 (2002)
13. Li, X.: A non-dominated sorting particle swarm optimizer for multiobjective optimization. In: Cantú-Paz, E., et al. (eds.) GECCO 2003. LNCS, vol. 2723, pp. 37–48. Springer, Heidelberg (2003)
14. Coello, C.A., Pulido, G.T., Lechuga, M.S.: Handling multiple objectives with particle swarm optimization. IEEE Trans. Evol. Comput. 8(3), 256–279 (2004)
15. Sierra, M.R., Coello, C.C.A.: Improving PSO-based multi-objective optimization using crowding, mutation and e-dominance. In: Coello, C.C.A., et al. (eds.) EMO 2005. LNCS, vol. 3410, pp. 505–519. Springer, Heidelberg (2005)
16. Hassan, R., Cohanim, B., Weck, O.: A comparison of particle swarm optimization and the genetic algorithm. In: Proceedings of 46th AIAA Structures, Structural Dynamics and Materials Conference, Austin, Texas (2005)
17. Niu, Y.F., Shen, L.C.: Multiobjective Constriction Particle Swarm Optimization and Its Performance Evaluation. In: Huang, D.S. (ed.) ICIC 2007. LNCS (LNAI), vol. 4682, pp. 1131–1140. Springer, Heidelberg (2007)
18. Graps, A.: An Introduction to Wavelets. IEEE Computational Science and Engineering 02(2), 50–61 (1995)
19. Mallat, S.: A Theory of Multiresolution Signal Decomposition: the Wavelet Representation. IEEE Trans. PAMI 11(7), 674–693 (1989)
20. Huang, X.S., Chen, Z.: A wavelet-based scene image fusion algorithm. In: Proceedings of IEEE TENCON 2002, Beijing, pp. 602–605 (2002)

21. Wang, Z., Bovik, A.C.: A universal image quality index. IEEE Signal Process Lett. 9(3), 81–84 (2002)
22. Wang, Z., Bovik, A.C., et al.: Image quality assessment: from error visibility to structural similarity. IEEE Trans. Image Process 13(4), 600–612 (2004)
23. Qu, G.H., Zhang, D.L., Yan, P.F.: Information measure for performance of image fusion. Electron Lett. 38(7), 313–315 (2002)
24. Ramesh, C., Ranjith, T.: Fusion performance measures and a lifting wavelet transform based algorithm for image fusion. In: Proceedings of FUSION 2002, Annapolis, pp. 317–320 (2002)
25. Wang, Q., Shen, Y., Zhang, Y., et al.: Fast quantitative correlation analysis and information deviation analysis for evaluating the performances of image fusion techniques. IEEE Trans. Instmm. Meas. 53(5), 1441–1447 (2004)
26. Tsagaris, V., Anastassopoulos, V.: A global measure for assessing image fusion methods. Opt. Eng. 45(2), 1–8 (2006)
27. Kennedy, J., Eberhart, R.C.: Particle swarm optimization. In: Proceedings of IEEE ICNN, Perth, pp. 1942–1948 (1995)
28. Kennedy, J., Eberhart, R.C.: Swarm Intelligence. Morgan Kaufmann, San Mateo (2001)
29. Ray, T., Liew, K.M.: A swarm metaphor for multiobjective design optimization. Eng. Optim. 34(2), 141–153 (2002)
30. Ho, S.L., Yang, S.Y., et al.: A particle swarm optimization-based method for multiobjective design optimizations. IEEE Trans. Magn. 41(5), 1756–1759 (2005)
31. Kennedy, J., Mendes, R.: Neighborhood Topologies in Fully Informed and Best-of-Neighborhood Particle Swarms. IEEE Trans. SMC Pt C: Appl. Rev. 4, 515–519 (2006)
32. Fang, K.T., Ma, C.X.: Orthogonal and uniform experimental design. Science Press, Beijing (2001)
33. Leung, Y.W., Wang, Y.P.: Multiobjective programming using uniform design and genetic algorithm. IEEE Trans. SMC Pt C Appl. Rev. 30(3), 293–304 (2000)
34. Reyes-Sierra, M., Coello, C.C.A.: Multi-objective Particle Swarm Optimizers: a Survey of the State-of-the-Art. Int. J. Comput. Intell. Res. 2(3), 287–308 (2006)
35. Bergh, F.V.D., Engelbrecht, A.P.: A Study of Particle Swarm Optimization Particle Trajectories. Information Sciences 176(8), 937–971 (2006)
36. Van Veldhuizen, D.A., Lamont, G.B.: Multiobjective Evolutionary Algorithm Research: A History and Analysis. Air Force Inst Technol, Wright-Patterson AFB, OH, Tech Rep. TR-98-03 (1998)
37. Czyzak, P., Jaszkiewicz, A.: Pareto-Simulated Annealing: A Metaheuristic Technique for Multi-Objective Combinatorial Optimization. Journal of Multi-Criteria Decision Analysis 7(1), 34–47 (1998)
38. Schott, J.R.: Fault Tolerant Design Using Single and Multicriteria Genetic Algorithm Optimization. M S Thesis, Massachusetts Inst. Technol., MA (1995)
39. Van Veldhuizen, D.A.: Multiobjective Evolutionary Algorithms: Classifications, Analyzes, and New Innovations. Ph D Dissertation, Graduate School of Eng., Air Force Inst. Technol., Wright-Patterson AFB, OH (1999)
40. Kita, H., Yabumoto, Y., et al.: Multi-objective Optimization by Means of The Thermodynamical Genetic Algorithm. In: Voigt, H.M., et al. (eds.) PPSN 1996. LNCS, vol. 1141. Springer, Heidelberg (1996)
41. Eberhart, R.C., Shi, Y.: Particle swarm optimization: development, applications and resources. In: Proceedings of IEEE CEC, Seoul, pp. 81–86 (2001)

8

Multi-objective Damage Identification Using Particle Swarm Optimization Techniques

Ricardo Perera and Sheng-En Fang

Department of Structural Mechanics, Technical University, Madrid 28006, Spain
perera@etsii.upm.es, shengen.fang@upm.es

The implementation of a technique that is able to detect the real state of a structure in near real time constitutes a key research field for guaranteeing the integrity of a structure and, therefore, for safeguarding human lives. This chapter presents particle swarm optimization-based strategies for multiobjective structural damage identification. Different variations of the conventional PSO based on evolutionary concepts are implemented for detecting the damage of a structure in a multiobjective framework.

8.1 Introduction

The implementation of a damage detection strategy for aerospace, civil and mechanical engineering infrastructures is referred to as structural health monitoring (SHM). Over the last few years, there have been increasing demands to develop SHM systems over different kinds of systems because of the huge economical and life-safety benefits that such technologies have the potential to provide.

Current damage detection methods are either visual or nondestructive experimental methods such as ultrasonic and acoustic emission methods, x-ray methods, etc. These kinds of experimental techniques are based on a local evaluation in easily accessible areas, and therefore, they require a certain a priori knowledge of the damage distribution. With the purpose of providing global damage detection methods applicable to complex structures, techniques based on modal testing [1]. and signal processing, constitute a promising approach for damage identification in civil, aeronautical and mechanical engineering. These methods examine changes in the dynamic characteristics of the structure, such as natural frequencies and mode shapes, etc, to detect the structural damage [2, 3]. The comparison between the undamaged and

N. Nedjah et al. (Eds.): Multi-Objective Swarm Intelligent Systems, SCI 261, pp. 179–207.
springerlink.com

damaged structure makes it possible to identify the location and the severity of damage.

One of the key points for the success of an SHM procedure is concerned with feature selection, i.e., the choice of the measurements taken as the basis for monitoring. Once the suitable physical parameters have been chosen, the comparison between the undamaged and damaged structure makes it possible to identify the location and severity of damage. The most usual approach for solving this sort of problem is the use of the finite element model updating method [4, 5, 6]. To apply the method, one objective function measuring the fit between measured and model-predicted data is chosen. Then, optimization techniques are used to find the optimal values of the model parameters that minimize the value of the objective function, i.e., those values best fitting the experimental data. Damage detection methods based on the model updating method have usually been developed as single objective optimization problems. However, with the lack of a clear objective function in the context of real-world damage detection problems, it is advisable to perform simultaneous optimizations of several objectives with the purpose of improving the performance of the procedure [7, 8]. Often, these objectives are conflicting. As opposed to single-objective optimization problems which accept one single optimum solution, multiobjective optimization problems do not have a single optimal solution, but rather a set of alternative solutions, named the Pareto front set, which are optimal in the sense that no other solutions in the search space are superior to them when all objectives are considered.

For dealing with multiobjective optimization problems, traditional gradient based optimizers use aggregating function approaches based on combining all the objectives into a single one using either addition, multiplication or any other combination of arithmetical operations by using some weighting factors. By varying the weights a set of Pareto optimal solutions is obtained, although it may not necessarily result in an even distribution of Pareto optimal points and an accurate, complete representation of the Pareto optimal set. Another problem with this method is that it is impossible to obtain points on non-convex portions of the Pareto optimal set in the criterion space.

Evolutionary algorithms (EA) seem to be particularly appropriate to these kinds of problems [9] because they search for a set of solutions in parallel in such a way that the search process can be driven towards a family of solutions representing the set of Pareto optimal solutions. Since the mid 1980s a considerable amount of research has been done in this area, especially in the last five years [10, 11] due to the difficulty of conventional optimization techniques being extended to multiobjective optimization problems. Because of this, several multiobjective EAs have been proposed in recent years [12]. However, evolutionary techniques require a relatively long time to obtain a Pareto front of high quality.

Particle swarm optimization (PSO) [13] is one of the newest techniques within the family of optimization algorithms and is based on an analogy with the choreography of flight of a flock of birds. The PSO algorithm relies only on two simple PSO self-updating equations whose purpose is to try to emulate the best global individual found, as well as the best solutions found by each individual particle. Since an individual obtains useful information only from the local and global optimal individuals, it converges to the best solution quickly. PSO has become very popular because of its simplicity and convergence speed and has been successfully implemented in various optimization problems like weight training in neural networks [14], functions optimization [15, 16, 17] and feature selection [18].

However, although several extensions to the PSO for handling multiple objectives have been proposed [19, 20, 21, 22, 23, 24], due to its single-point-centered characteristic, the conventional PSO does not perform well in real-world complex problems like those of damage identification in which the search has to be made in multi-constrained solution spaces. The location of the non-dominated points on the Pareto front will be difficult since more than one criterion will direct the velocity and position of an individual. Because of this, there are many associated problems that require further study for extending PSO in solving multi-objective problems. For example, although the sharing of information among particles based on their previous experiences contributes to increasing the convergence speed, it can be a demerit in multiobjective problems since it reduces the diversity of the algorithm.

Despite the fact that EA has been widely applied to solve damage detection problems [7, 8, 25], PSO has never been considered in the context of these problems although its potential use, because of its high convergence speed and feasibility of implementation, makes it an ideal candidate to implement a structural health monitoring technique . For this reason, the goal of this paper is to present the first application of PSO to multiobjective damage identification problems and investigate the applicability of several variations of the basic PSO technique. The potential of combining evolutionary computation and PSO concepts for damage identification problems has never been performed before and is explored in this work by using a multiobjective evolutionary particle swarm optimization algorithm [MOEPSO]. A successful combination would make MOEPSO an ideal candidate to develop a structural health monitoring system in real time for complex structures since it would combine the high convergence speed of PSO with the feasibility of EAs to solve multiobjective problems.

This work is organized as follows. Sections 8.2 and 8.3 describe the general concepts involved in the modelling of multiobjective damage identification problems. Some background information about PSO is given in Section 8.4, while the details of the proposed MOEPSO are described in Section 8.5. The performance of the proposed algorithms on numerical and experimental damage identification problems is shown in Section 8.6. The effects of the proposed features are examined and a comparative study of the proposed MOEPSO with well-known multiobjective optimization EAs is presented. Conclusions are drawn in Section 8.7.

8.2 Single Objective Damage Identification Problem Formulation

Most of the approaches used for damage identification are based on methods of model updating which are posed as minimization problems of a single objective function which usually consists of the error between the measured vibration data and the corresponding predictions of a finite element (FE) numerical model. Parameters able to identify the structural damage are then chosen that are assumed uncertain, and these are estimated by using an iterative optimization process because of the nonlinear relation between the vibration data and the physical parameters.

Therefore, setting-up an objective function, selecting updating parameters and using robust optimization algorithms are three crucial steps in finite element model updating procedure.

The objective function is usually expressed in the following way

$$F(\boldsymbol{\theta}) = ||\mathbf{z}_m - \mathbf{z}(\boldsymbol{\theta})||^2 \tag{8.1}$$

where $\mathbf{z}_m$ and $\mathbf{z}(\boldsymbol{\theta})$ are the measured and computed modal vectors (usually natural frequencies and less often mode shapes) and $\boldsymbol{\theta}$ is a vector of all unknown parameters which should be indicative of the level of structural damage. The objective function in Eq.(8.1) is a nonlinear function of the parameters $\boldsymbol{\theta}$ and its minimization would allow determining the value of the parameters $\boldsymbol{\theta}$, i.e. the level of damage for which the numerical results $\mathbf{z}(\boldsymbol{\theta})$ coincide with the measured values $\mathbf{z}_m$. The minimization may be performed using methods such as sensitivity-based methods, quadratic programming, simulated annealing or genetic algorithms.

The third key aspect of a model-based identification method is the choice of the unknown parameters $\boldsymbol{\theta}$, i.e. the parameterization of the candidate damage. Since inverse approaches rely on a model of the damage, the success of the estimation is dependent on the quality of the model used. The type of model used will depend on the type of structure and the damage mechanism and, according to these, may be simple or complex. Using a measured modal model consisting of the lower natural frequencies and associated mode shapes will mean that only a coarse model of the damage may be identified although sufficient for a first estimation of damage in the implementation of a structural health monitoring procedure.

The effect of damage on the finite element formulation is introduced according to Continuum Damage Mechanics; in order to do this, damage is quantified through a scalar variable or index d whose values are between 0 and 1. A zero value corresponds to no damage while values next to one imply rupture. In the context of discretized finite elements, the definition of a damage index d_e for each element allows estimating not only the damage severity but also the damage location since damage identification is then carried out at the element level. Since it has been assumed that no alteration occurs before and after damage related to the mass, which is acceptable in most real applications, damage indices will be incorporated in the finite element formulation of the studied problem through the stiffness of each element. As in our work, single beam finite elements have been used to represent the structure, the parameterization of the damage has been represented by a reduction factor or damage index of the element bending stiffness. This damage index represents the relative variation of the element bending stiffness, $(EI)^e_d$, to the initial value, $(EI)^e$:

$$d_e = 1 - \frac{(EI)^e_d}{(EI)^e} \tag{8.2}$$

where E^e and I^e are the Youngs modulus and the inertia of each element, respectively. The damage indices d_e are the parameters to be estimated during the updating procedure with the purpose of determining the real state of the different parts of the structure. In Eq.(8.1), the $\boldsymbol{\theta}$ vector would coincide with the vector grouping the damage indices of all the elements of the finite element mesh. When $d_e = 1$ in Eq.(8.2) $(EI)^e_d$ is equal to zero, i.e., the stiffness of the element e is zero what means it is fully damaged; However, if $d_e = 0$ then the actual stiffness of the element $(EI)^e_d$ coincides with the initial stiffness $(EI)^e$ what means that the element is not damaged.

Damage identification methodology with EAs proceeds using the following steps: (a) At the outset of the optimization of the geometry, the boundaries and the material properties of the structure under inspection have to be defined. In the same way, the frequencies and mode shapes of the structure experimentally tested are provided since modal parameters are representative of the real state of the structure. (b) The design variables are properly selected. In our case these variables are the element bending stiffness reduction factors or damage indices. (c) Using finite elements, the frequencies and mode shapes of the numerical model are evaluated. Since the modal parameters are dependent on the structural stiffness, their value will depend on the damage indices. (d) The damage indices are updated using the selection, crossover and mutation operators with the purpose of minimizing an objective function, measuring the difference between the experimental and numerical modal parameters. (e) If the convergence criteria for the optimization algorithm are satisfied, then the optimum solution, i.e. the real damage of the structure, has been found and the process is terminated; if not the optimizer updates the design variable values and the whole process is repeated from Step (c).

8.3 Multi-objective Damage Identification

8.3.1 Formulation of the Multi-objective Problem

Most damage detection problems are treated with a single objective. However, in real-world applications, a single objective function is rarely sufficiently representative of the performance of the structure. Because of this, it is advisable to consider simultaneous optimizations of several objectives with the purpose of improving the robustness and performance of the procedure. In fact, nowadays, multicriteria optimization has begun to be applied to the damage identification of aeronautical, mechanical and civil engineering structures.

In general, the mathematical formulation of a multiobjective problem can be defined as follows. To find the values of the damage parameter set **d** that simultaneously minimizes the objectives

$$F(\mathbf{d}) = (F_1(\mathbf{d}), F_2(\mathbf{d}), \ldots, F_m(\mathbf{d})) \tag{8.3}$$

where $\mathbf{d} = (d_1, d_2, \ldots, d_{NE})$ is the damage vector reflecting the damage value for each one of the NE finite elements of the structure. Furthermore, each one of the damage variables is confined to satisfy the following constraint

$$0 \leq \mathbf{d}_i \leq 1 \quad i = 1, 2, \ldots, NE \tag{8.4}$$

Therefore, the problem is defined by m objective functions to be minimized and NE design variables or optimization parameters.

The predominant solution concept in defining solutions for multiobjective optimization problems is that of Pareto optimality. To define this concept mathematically we assume, without loss of generality, a minimization problem and consider two decision vectors **d** and $\mathbf{d}^*$. Then, $\mathbf{d}^*$ is said to be nondominated if there is no other vector **d** in the search space, such that:

$$F_i(\mathbf{d}) \leq F_i(\mathbf{d}^*) \quad \forall i = 1, \ldots, m \tag{8.5}$$

$$\text{with} \quad F_j(\mathbf{d}) < F_j(\mathbf{d}^*) \quad \text{for at least one objective j} \tag{8.6}$$

The non-dominated solutions are optimal in the sense that no other solutions in the search space are superior to them when all objectives are considered and are denoted as Pareto optimal. The set of Pareto optimal constitutes the so-called Pareto-optimal set or Pareto-optimal front.

Closely related to Pareto optimality, a vector $\mathbf{d}^*$ is weakly Pareto optimal if and only if there is no other vector **d** in the search space, such that:

$$F_i(\mathbf{d}) < F_i(\mathbf{d}^*) \quad \forall i = 1, \ldots, m \tag{8.7}$$

That is, there is no vector that improves all the objective functions simultaneously. In contrast to weakly Pareto optimal points, no objective function can be improved from a Pareto optimal point without detriment to another objective function.

The goal of multiobjetive optimization is to find the global Pareto optimal set, although from a practical point of view, a compromise should be made among the available Pareto solutions to find a single final solution or a set of representative solutions.

8.3.2 Objective Functions

In formulating a damage detection problem the choice of the objective functions (Eq.(8.1)) represents the most important decision to be performed. In fact, many different single objective functions, depending directly or indirectly on basic modal parameters, have been proposed in recent years [5, 25, 26, 27]. However, there is not a clear criterion for choosing the suitable objective function. Due to this, a combined consideration of some of them can be a good solution.

Among the various choices available, in this work two different objective functions have been adopted to perform multiobjective damage identification.

The first objective function contains all modal frequencies with the measure of fit selected to represent the mismatch between the measured and the model predicted frequencies for all modes. It has been formulated as follows:

$$F_1 = \frac{1}{N_m} \sum_{j=1}^{N_m} \frac{||\omega_{j,num}(d) - \omega_{j,exp}||^2}{||\omega_{j,exp}||^2} \tag{8.8}$$

where ω are the modal frequencies and N_m is the number of experimentally identified modes. Subscripts num and exp refer to the numerical and experimental results, respectively. Function F_1, such as defined in Eq.(8.8), is normalized between 0 and 1.

The second objective function contains the mode shape components for all modes with the measure of fit selected to represent the mismatch between the measured and the model predicted mode shape components for all modes. To correlate experimental and numerical mode shapes, the modal assurance criterion (MAC) [1] defined as

$$MAC(\{\phi_{j,exp}\}, \{\phi_{j,num}\}) = \frac{|\{\phi_{j,exp}\}^t \{\phi_{j,num}\}|^2}{(\{\phi_{j,exp}\}^t \{\phi_{j,exp}\})(\{\phi_{j,num}\}^t \{\phi_{j,num}\})} \tag{8.9}$$

has been used. In Eq.(8.9) ϕ_j is the jth mode shape.

The MAC criterion measures the correlation between two vectors, in our case one calculated, $\phi_{j,num}$, and the other experimentally measured, $\phi_{j,exp}$. In general terms, the MAC provides a measure of the least-squares deviation or scatter of the points from the straight line correlation and its values are between 0 and 1, which allows formulating the objective function in a normalized way. Low values mean low correlation between the vectors while high values indicate a high correlation.

From Eq.(8.9), the second objective function is defined considering all the measured modes as follows

$$F_2 = 1 - \prod_{j=1}^{N_m} MAC(\{\phi_{j,exp}\}, \{\phi_{j,num}\}) \tag{8.10}$$

The use of functions F_1 and F_2 allows estimating all Pareto optimal models that trade-off the overall fit in modal frequencies between the overall fit in the mode shapes.

8.4 Overview of Basic Particle Swarm Optimization (PSO)

Particle swarm optimization is a technique first originated by Kennedy and Eberhart [13] from the social behavior of bird flocking and fish schooling. In the same way as evolutionary algorithms, PSO is initialized with a swarm of random particles and then, using an iterative procedure the optimum is searched. For every updating cycle, each particle is updated such that it tries to emulate the global best particle, known as *gbest*, found so far in the swarm of particles, and the best solution, known as *pbest*, found so far by particle i, i.e., the number of *pbest* particles agrees with the number of particles in the swarm. To perform this, self-updating equations are used as follows:

$$v_i = w \cdot v_i + c_1 \cdot r_1 \cdot (pbest_i - x_i) + c_2 \cdot r_2 \cdot (gbest - x_i) \tag{8.11}$$

$$x_i = x_i + v_i \tag{8.12}$$

where v_i is the particle velocity, x_i is the current position of particle i, w is an inertia coefficient balancing global and local search, r_1 and r_2 are random numbers in [0,1] and c_1 and c_2 are the learning factors which control the influence of $pbest_i$ and $gbest$ on the search process. Usually, values equal to 2 are suggested for c_1 and c_2 for the sake of convergence [16]. Additionally, the velocity is limited to a maximum value with the purpose of controlling the global exploration ability of particle swarm avoiding it moving towards infinity. Eqs.(8.11) and (8.12) represent the original PSO algorithm although some variations have been proposed [28].

The inertia weight w is an important factor for the PSOs convergence. It controls the impact of the previous history of velocities on the current velocity. A large inertia weight factor facilitates global exploration while a small weight factor facilitates local exploration. Therefore, it is advisable to choose a large weight factor for initial iterations and gradually reduce the weight factor in successive iterations. This can be done by using

$$w = w_{max} - \frac{w_{max} - w_{min}}{iter_{max}} \cdot iter \tag{8.13}$$

where w_{max} is the initial weight, w_{min} is the final weight, $iter_{max}$ is the maximum iteration number, and *iter* is the current iteration number.

From a psychological point of view, the first term in Eq.(8.11) is related to the movement inertia of the particles, the second term is the cognitive term whose purpose is to try to duplicate successful past behaviors of the particle and the last term, the social term, represents the tendency to imitate the successes of other particles.

Normally, PSO is considered as an evolutionary approach since it possesses many common characteristics used by EAs, such as: a) It is initialized with a randomly generated population. b) The optimum is searched by updating generations. c) Fitness evaluation is evaluated by objective functions. However, unlike EAs, crossover and mutation operators are not applied.

8.5 Multi-objective PSO

To extend the basic PSO algorithm to solve multiobjective problems two main issues should be addressed according to the Pareto dominance concept: (a) How to assign the fitness value and how to select the best particles with the purpose of extending the existing particle updating strategy in PSO to account for the requirements in multiobjective optimization .(b) In contrast to single objective optimization, it is essential to obtain a well-distributed and diverse solution set for finding the final tradeoff in multiobjective optimization. Because the search procedure by the PSO depends strongly on *pbest* and *gbest*, the search can be limited by them, i.e. diversity is introduced only around the elite individuals found through the random elements in the PSO equations. Therefore, the choice of the means to ensure the diversity of the Pareto front constitutes another essential point for attaining the success of the procedure.

Based on the similarity between PSOs and EAs, these two issues might be considered by incorporating multiobjective handling techniques available in EAs [23]. An optimal modification of the basic PSO for multiobjective optimization should keep its main advantages and, furthermore, should preserve population diversity for finding the optimal Pareto front. Using a combination between PSO and EA, an efficient and effective multi-objective algorithm might result by integrating the main advantages of each one.

In this section, detailed implementation of the proposed multiobjective evolutionary particle swarm optimization algorithm [MOEPSO] is given. In the proposed method, a truncated elite archive is maintained which contains a set of nondominated solutions. New nondominated solutions are included in the archive in each iteration, and the archive is updated to make it domination free. Furthermore, typical genetic operators like mutation and crossover are included. The flowchart of the MOEPSO is shown in Fig.8.1. Next, the main features of the proposed algorithm are presented.

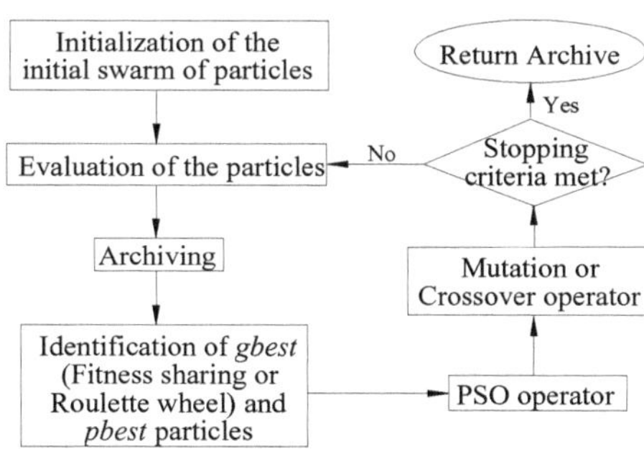

Fig. 8.1. Flowchart of MOEPSO for solving damage identification problems

8.5.1 Initialization and Constraint Handling

According to the formulation of the problem in Sections 8.2 and 8.3, for damage identification the only constraint is related to the value of the damage indices representing the location and severity of the damage at each element of the FE mesh. From a physical point of view, these indices should have values between 0 and 1 or, more suitably, between 0 and 0.99 to avoid numerical problems. However, by applying Eqs (9.11) and (9.12) any value would be possible which would contribute hugely to decreasing the applicability of PSO for solving this kind of problem. To avoid this, a simple but effective method has been used to solve the constrained optimization problem. On the one hand, during the initialization process, all particles are started with feasible solutions. On the other hand, during the updating procedure according to Eqs.(8.11) and (8.12), all the particles are kept within the search space, i.e. solutions which do not fall within the valid search space are not accepted. When a decision variable goes beyond its boundaries, it is forced to take the value of the corresponding boundary (either the lower or upper boundary).

Compared to other constraint handling techniques, this scheme is simple, fast and easy to apply since any complicated manipulation is not required; fitness functions and constraints are handled in a separate way.

8.5.2 Archiving

In MOEPSO, the choice of *gbest* in Eq.(8.11) plays a very important role in guiding the entire swarm towards the global Pareto front. However, unlike single objective optimization, the *gbest* for multiobjective optimization exists in the form of a set of nondominated solutions. Therefore, elitism is implemented in the form of an external fixed-size archive or repository in which a historical record of the nondominated particles found along the search process is stored. The archive is updated at each generation, e.g., if the candidate particle is not dominated by any members in the archive, it will be added to the archive. Likewise, any archive members dominated by this solution will be removed from the archive.

Too many nondominated solutions in the external repository would not be convenient since they might slow down the search and, furthermore, would be useless when exceeding a reasonable number. Therefore, when the archive has reached its maximum permitted capacity, a recurrent truncation process is performed to remove some particles from the repository. On the other hand, if the points in the external repository are not distributed uniformly, the fitness assignment method might be possibly biased toward certain regions of the search space, leading to an unbalanced distribution in the population. The average linkage method [29] has been chosen here to prune the repository while maintaining its characteristics. The basic idea of this method is the division of the nondominated particles into groups of relatively homogeneous elements according to their distance. The distance d between two groups, g_1

and g_2, is given as the average distance between pairs of individuals across the two groups

$$d = \frac{1}{|g_1| \cdot |g_2|} \cdot \sum_{i_1 \in g_1, i_2 \in g_2} ||i_1 - i_2|| \tag{8.14}$$

where the metric $||\cdot||$ reflects the distance between two individuals i_1 and i_2.

Then, following an iterative process, the two groups or clusters with minimal average distance are amalgamated into a larger group until the number of clusters is equal to the maximum permitted capacity of the repository. Finally, the reduced nondominated set is computed by selecting a representative individual per cluster, usually the centroid. With this approach a uniform distribution of the grid defined by the nondominated solutions can be reached.

8.5.3 Selection of *pbest* and *gbest*

The selection of social and cognitive leaders, *gbest* and *pbest*, plays a critical role in MOPSO algorithms. The selection of the cognitive leader follows the same rule as in traditional PSO. The only difference is that Pareto dominance is applied, i.e., if the current particle is dominated by its better position until now stored in memory, then the position in memory is kept. Otherwise, the current position replaces the one in memory; if neither of them is dominated by the other, the best position is selected randomly.

In order to promote diversity and to encourage exploration of the least populated region in the search space, two selection criteria of *gbest* for reproduction have been employed:

(a) Niche count: From all the particles of the external repository, the choice is performed by fitness sharing [30], i.e., the individual of the sparsest population density in its niche is selected, the size of the niche being determined by the niche radius, σ_{share}, whose value is dependent on the problem. The main purpose of this approach is to promote diversity and to encourage exploration of the least populated region in the search space by degrading those individuals that are represented in the higher percentage of the population.

(b) Roulette wheel selection scheme based on fitness assignment: All the candidates are assigned weights based on their fitness values; then the choice is performed using roulette wheel selection. The fitness assignment mechanism for the external population proposed by Zitzler and Thiele [31] has been used in the proposed algorithm. According to this mechanism each individual i in the external repository is assigned a strength s_i proportional to the number of swarm members which are dominated by it, i.e.

$$s_i = \frac{n_i}{N+1} \tag{8.15}$$

where n_i is the number of swarm members which are dominated by individual i and N is the size of the swarm. The fitness value of the individual i is the inverse of its strength. Therefore, nondominated individuals with a high strength value and, therefore, populated densely are penalized and fitness sharing is not needed.

8.5.4 Use of a Mutation Operator

PSO is known to have a very high convergence speed. However, in the context of complex multiobjective optimization problems, which are usual for many real world problems, such convergence speed may result in a premature convergence to a false Pareto front due to the difficulty of incorporating the heuristic characteristic of complex problems in the algorithm by using only the two PSO equations. This phenomenon might be avoided by encouraging the exploration to regions beyond those defined by the search trajectory. This is the main motivation for using EA operators.

Mutation operators have usually been applied. This operator might be developed using a self-adaptive mechanism in such a way that high exploratory capabilities are imparted to the particles at various stages of the search by varying the probability of mutation with the requirements of the search. Coello *et al* [23] proposed an operator with a high explorative behaviour at the beginning decreasing its effect as the number of iterations increases. Agrawal *et al* [32] proposed a self-adaptive operator with varying probabilities of mutation according to the number of particles in the external repository. Alternatively, the number of particles affected by the mutation operator might remain almost constant through the iterative procedure by defining a constant mutation probability. In this way, the exploratory ability remains constant along the iterative procedure. Generally, a low value is usually chosen for the probability of mutation with the purpose of not increasing excessively the diversity of the population.

On the other hand, the periodical use of a crossover operator might help to increase the exploration ability in searching solution space by periodically changing the internal structure of the particles by sharing information with other different individuals apart from the best individuals. Therefore, the crossover does not create new material within the population; it simply intermixes the existing population. For its application, two particles are randomly selected and then one crossover point is randomly selected. Taking this point as reference, a partial swap mechanism is performed between the two selected individuals. The probability of crossover defines the ratio of the number of offspring produced in each generation to the population size. Taking into account that this operator is not an essential operator for the procedure but a complement to wide the explored region a not very high value should be taken for the probability of crossover.

8.6 Benchmarking Experiments

Two different damage identification problems solved using the objective functions in Eqs.(8.8) and (8.10) were taken to compare the proposed approaches. The first of them corresponds to a simply supported beam damaged in two different sections with two different severities while the second one is a damage detection problem on a reinforced concrete beam.

Six versions of the PSO algorithm are compared to illustrate the individual and combined effects of the proposed features. These versions include standard MOPSO, MOEPSO with mutation operator (MOEPSO-M) and MOEPSO with crossover operator (MOEPSO-C). In the three cases, selection schemes by fitness sharing (FS) and by roulette wheel were considered (RW). For fitness sharing a niche radius of 0.1 was adopted by the previous experience of the authors in the application of genetic algorithms based on fitness sharing to damage identification problems. In the same way, in all the studies performed, a crossover probability of 0.1 has been assumed for the MOEPSO-C and a constant mutation probability of 0.01 for the MOEPSO-M. The value of the probability of the mutation is very usual in the application of genetic algorithms. High values for this parameter might decrease the speed of convergence considerably. On the other hand, in the application of genetic algorithms the probability of crossover is usually chosen higher than 0.5 since it is a basic operator in the algorithm. This is not the case in the proposed algorithm in this work. Because of it, after some numerical tests a value of 0.1 was chosen to give good performance although values slightly higher might be also suitable. A size of 50 was adopted for the external repository.

With the purpose of checking the performance of the proposed multiobjective procedure, typical values were selected for the PSO parameters of Eqs.(8.11), (8.12) and (8.13): Cognitive parameter $c_1 = 2$, social parameter $c_2 = 2$, initial inertia weight $w_{max} = 0.95$, final inertia weight $w_{end} = 0.4$, maximum velocity $v_{max} = 100$.

All the simulations were performed on an Intel Pentium 3 GHz personal computer. Thirty and one hundred simulation runs were performed for each algorithm in order to study the statistical performance. A random initial population was created for each of the 30 and 100 runs.

The comparisons were made from results derived after an identical number of objective function evaluations were performed, using that factor as a measure of common computational effort. As is known, a good representation of the Pareto frontier involves diversity and precision of the optimal solution set. Then, the comparison of the above MO algorithms for damage detection purposes can be focused on it.

When dealing with multiobjective optimization problems, a qualitative assessment of results becomes difficult because several solutions will be generated instead of only one. Different performance measures or indicators can be used to evaluate the Pareto fronts produced by the various algorithms. These include the size of the Pareto front, the average individual density value and the coverage of the different Pareto fronts. This last measure allows setting if the outcomes of one algorithm dominate or not the outcomes of another algorithm.

8.6.1 Simply Supported Beam

The first test problem for a comparative investigation consists in identifying damage for a simply supported concrete beam of length $L = 5$ m and rectangular cross section $b \times h = 0.25\text{m} \times 0.2\text{m}$. The beam was assumed to have a bending stiffness EI of 5000000 Nm^2 and a density ρ of 2500 kg/m^3.

The beam was subjected to a multiple simulated damage scenario (Fig. 8.2) of complex identification. The "measured"modal responses of the beam before and after damage were generated numerically by using a finite element as shown in Fig.8.2. In spite of the modal values being numerically simulated, to be more consistent with the field test conditions and to check the robustness of the proposed procedures, only the four lowest vibration modes were considered, and due to the limited number of sensors existing in the real practice, the mode shape vector was only read at a limited number of locations coincident with the vertical degrees of freedom of the nodes in Fig. 8.2. Furthermore, to test the stability and robustness of the proposed algorithms, a random artificial noise of level equal to 15% was added to the theoretically calculated frequencies and mode shapes. The addition of this severe noise contributes to increase the difficulty of the detection procedure.

The numerically generated measurements, instead of experimental measurements, were used to check the performance of the six versions of PSO algorithm specified above.

The swarm size was fixed to 50 particles and as stop criterion the total number of objective function evaluations was set at 10000×30 and 10000×100 for all the approaches depending on the number of runs, i.e., 100 iterations per run. A relatively small number of evaluations were set to examine the convergence of proposed algorithms.

With the conditions mentioned above, the comparison of the Pareto front curves for the six versions of PSO are shown in Figs. 8.3, 8.4 and 8.5, with

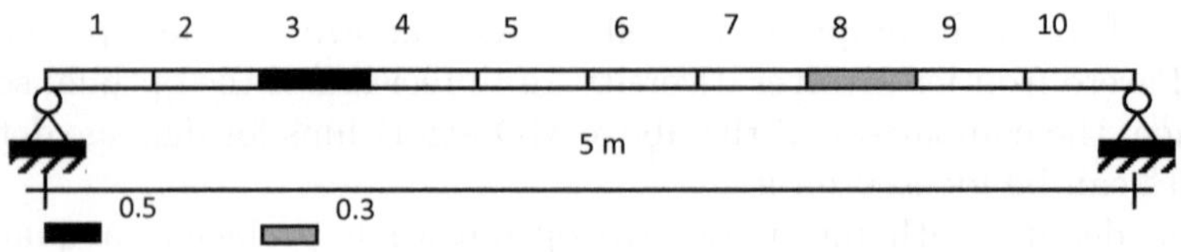

Fig. 8.2. Finite element mesh and damage scenario for the beam

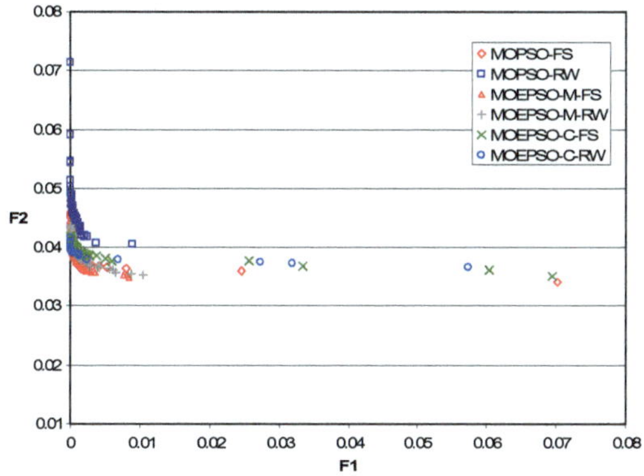

(a)

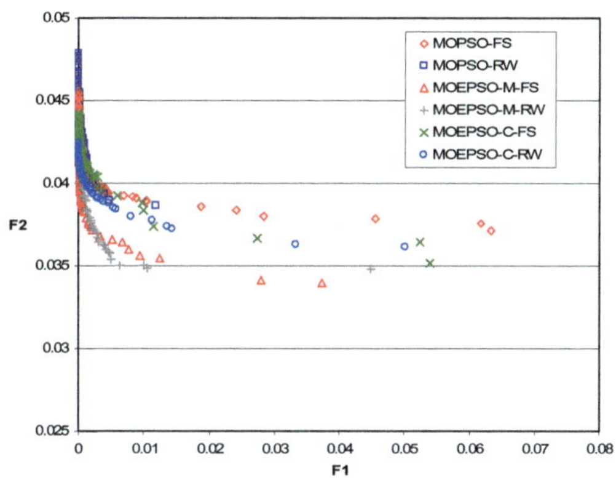

(b)

Fig. 8.3. Pareto fronts for the six versions of the PSO algorithm: Beam. (a) 30 runs; (b) 100 runs

the objective functions F_1 and F_2 on the horizontal and vertical axis, respectively. The graphical representations indicate, at first glance, that, in general, MOEPSO-M performs very well in comparison with its competitors. Furthermore, selection scheme based on fitness sharing performs better than that based on roulette wheel.

Pareto front size and average density values over 30 and 100 runs by the six algorithms are shown in Table 8.1 and 8.2. The improvement in the values when the number of runs increases is evident.

Table 8.3 and 8.4 compares the six algorithms in pairs using values for the C metric defined in Zitzler and Thiele [31]. The C metric measures the coverage of two sets of solution vectors. Let X' and $X'' \subseteq X$ be two sets

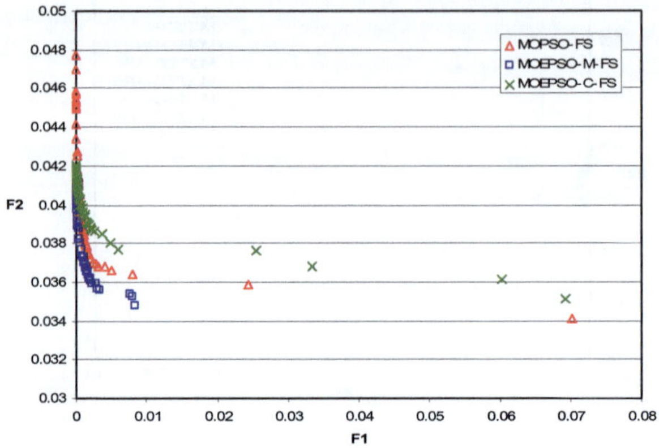

(a)

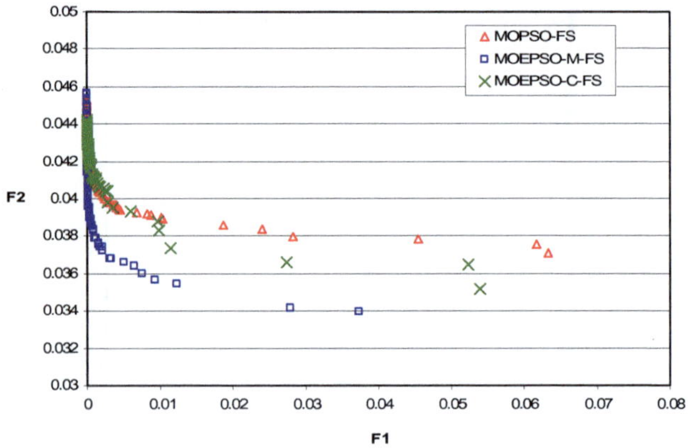

(b)

Fig. 8.4. Pareto fronts for the three versions of the PSO algorithm based on FS: Beam. (a) 30 runs; (b) 100 runs

Table 8.1. Average density values and Pareto front sizes: Beam (30 runs)

	MOPSO-FS	MOPSO-RW	MOEPSO-M-FS	MOEPSO-M-RW	MOEPSO-C-FS	MOEPSO-C-RW
Average Density Value	10.23	8.62	13.17	9.18	16.35	14.23
Pareto front size	59	44	49	48	52	35

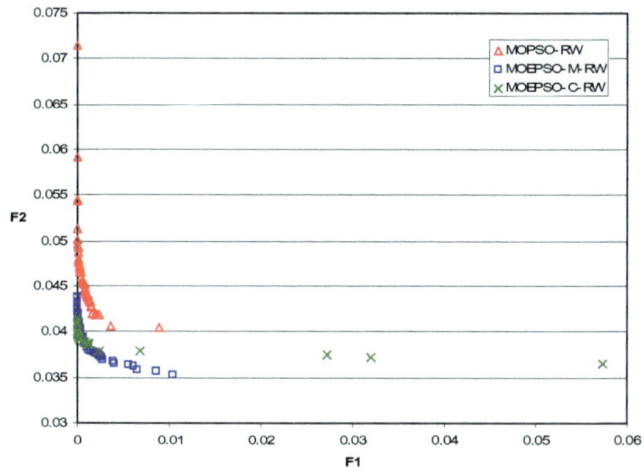

(a)

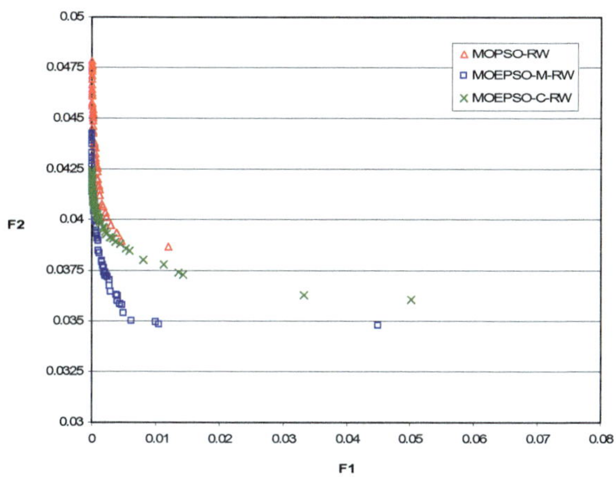

(b)

Fig. 8.5. Pareto fronts for the three versions of the PSO algorithm based on RW: Beam. (a) 30 runs; (b) 100 runs

Table 8.2. Average density values and Pareto front sizes: Beam (100 runs)

	MOPSO-FS	MOPSO-RW	MOEPSO-M-FS	MOEPSO-M-RW	MOEPSO-C-FS	MOEPSO-C-RW
Average Density Value	29.65	10.82	19.12	11.49	30.17	27.81
Pareto front size	35	55	102	73	85	79

Table 8.3. C metrics to measure the coverage of two sets of solutions: Beam. X′: First column; X″: First line (30 runs)

	MOPSO-FS	MOPSO-RW	MOEPSO-M-FS	MOEPSO-M-RW	MOEPSO-C-FS	MOEPSO-C-RW
MOPSO-FS	N/A	0.93	0	0.2	0.43	0.19
MOPSO-RW	0	N/A	0	0	0	0
MOEPSO-M-FS	0.98	1	N/A	1	1	0.47
MOEPSO-M-RW	0.8	0.98	0	N/A	0.62	0.25
MOEPSO-C-FS	0.52	0.98	0	0.35	N/A	0.03
MOEPSO-C-RW	0.65	0.98	0.24	0.53	0.89	N/A

Table 8.4. C metrics to measure the coverage of two sets of solutions: Beam. X′: First column; X″: First line (100 runs)

	MOPSO-FS	MOPSO-RW	MOEPSO-M-FS	MOEPSO-M-RW	MOEPSO-C-FS	MOEPSO-C-RW
MOPSO-FS	N/A	0.91	0	0	0.12	0
MOPSO-RW	0.08	N/A	0	0	0.06	0
MOEPSO-M-FS	0.88	0.98	N/A	0.67	0.7	0.68
MOEPSO-M-RW	0.91	0.98	0.15	N/A	0.65	0.42
MOEPSO-C-FS	0.66	0.85	0.1	0.11	N/A	0
MOEPSO-C-RW	0.91	0.98	0.27	0.37	0.9	N/A

of solution vectors. The function C maps the ordered pair (X', X'') to the interval [0, 1]

$$C(X', X'') = \frac{|\{a'' \in X''; \exists a' \in X' : a' \succeq a''|}{|X''|} \tag{8.16}$$

The value $C(X', X'') = 1$ means that all the points in X'' are dominated by or equal to points in X'. The opposite, $C(X'', X') = 0$, represents the situation when none of the points in X'' is weakly dominated by X'. Both,

$C(X', X'')$ and $C(X'', X')$, have to be considered, since $C(X', X'')$ is not necessarily equal to $1 - C(X'', X')$. The C metric values used in the Table 8.3 and 8.4 indicate that MOEPSO-M-FS seems to provide the best performance. In general terms, the introduction of the mutation operator contributes to improve the results. The improvement with the crossover operator is evident in case of using a roulette wheel scheme but not so much when fitness sharing selection is applied.

Comparative study with EAs

In this section, the performance of MOEPSO is compared to two existing multiobjective EAs, NPGA [33] and SPGA [31]. The parameter settings of the two algorithms are shown in Table 8.5; the chosen parameters are standard for both algorithms. Thirty runs were also performed and the same stop criterion used previously was kept.

SPGA and NPGA Pareto fronts are shown in Fig. 8.6 where by clarity only MOEPSO-M-FS has been plotted of the PSO based algorithms. From this figure, we can see the advantage of PSO-based algorithms over EAs in MO problems in that the first produce a much faster convergence speed. Since the number of function evaluations used as stop criterion is small for the EAs the quality of their Pareto front is much worse from any point of view. It is evident that EAs would require a higher number of iterations to produce Pareto fronts comparable to those obtained with PSO-based algorithms, which demonstrates that MOPSO and MOEPSO algorithms are a strong competitor to EAs in solving multiobjective problems. In PSO-based algorithms each particle knows its moving direction and how fast it should go if another individual exists with better performance. This means that the probability of generating a better-fitted offspring by MOPSO algorithms is higher than for EAs since the risk of being trapped by a local Pareto front is decreased. Therefore, less evaluation time and fewer generations would be necessary with MOPSO algorithms in order to converge to a uniformly distributed Pareto front.

Table 8.5. Parameters settings of SPGA and NPGA algorithms

Population	Population size 50, External archive size 20 in SPGA
Representation	30 bits per individual
Selection	Tournament selection scheme
Crossover probability	0.7 (one point crossover scheme)
Mutation probability	0.01
Niche radius	0.1 in NPGA
Domination pressure	10%

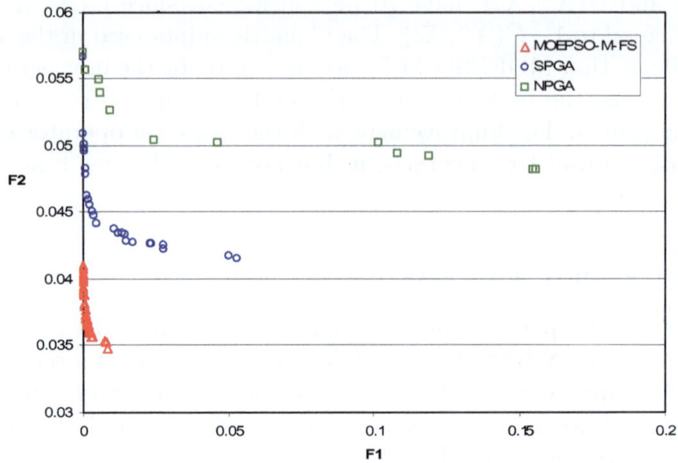

Fig. 8.6. Comparison of Pareto fronts among SPGA, NPGA and MOEPSO-M-FS

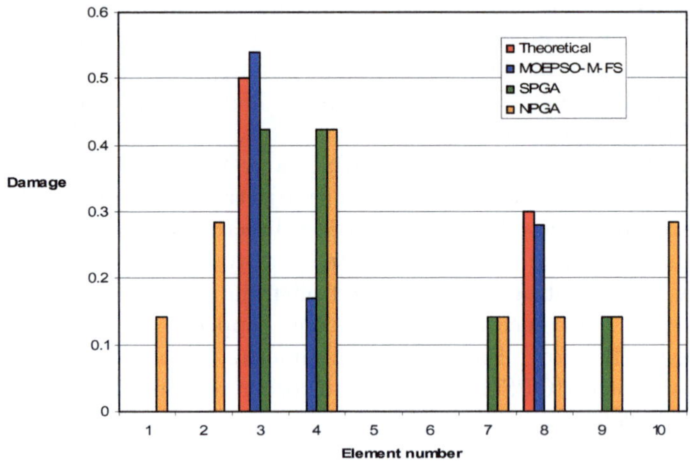

Fig. 8.7. Comparison of damage predictions among MOEPSO-M-FS, SPGA and NPGA for the beam

Finally, damage predictions for the simulated beam with MOEPSO-M-FS are compared to NPGA and SPGA and to the theoretical values shown in Fig. 8.7. To perform this objective, from the set of Pareto optimum solutions obtained for each algorithm, the chosen one is the one minimizing the following expression

$$\sqrt{F_1^2 + F_2^2} \tag{8.17}$$

In general, taking into account that noise was added, very good damage predictions were obtained with MOEPSO-M-FS. The same conclusion cannot be extended to EA predictions since a high number of false warnings were obtained in some elements. As commented above, a higher number of iterations should be performed to obtain more acceptable results, which again demonstrates the efficiency and effectiveness of PSO methods.

8.6.2 Experimental Reinforced Concrete Frame

The proposed methods were applied to estimate damage in a one-storey and one-bay RC frame experimentally tested in the Structures Laboratory of the Structural Mechanics Department of Madrid Technical University (Spain) (Fig.8.8). Unlike the previous example, the data used in this example were obtained through experimental tests, which contributes to increasing the difficulty of the identification procedure because of the uncertainties introduced in the tests. Furthermore, the procedure becomes more complex by considering concrete, since for this material damage appears as widespread cracking, which is more difficult to localize.

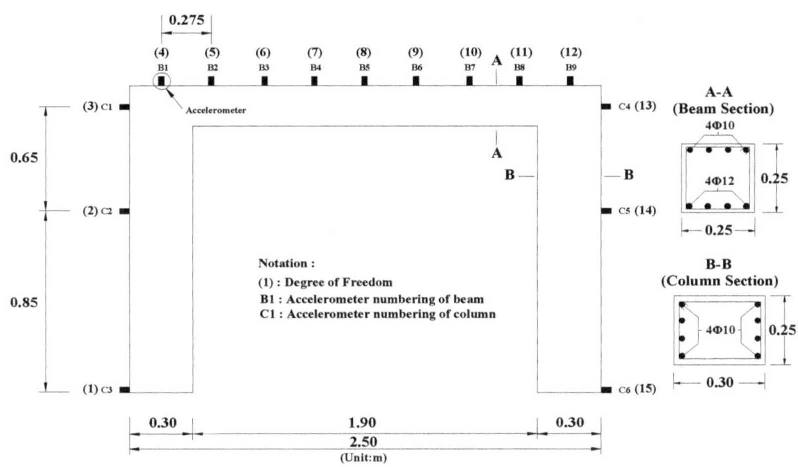

Fig. 8.8. Reinforced concrete frame experimentally tested

Damage was provided by applying an increasing static load. At the end of the loading process, damage appeared distributed at the mid-span of the frame and cracks were originated at the beam part of the joint. Modal tests were performed with the purpose of using their results to verify the performance of the multiobjective updating procedure developed in Sections 8.2 and 8.3 as a method for identifying damage.

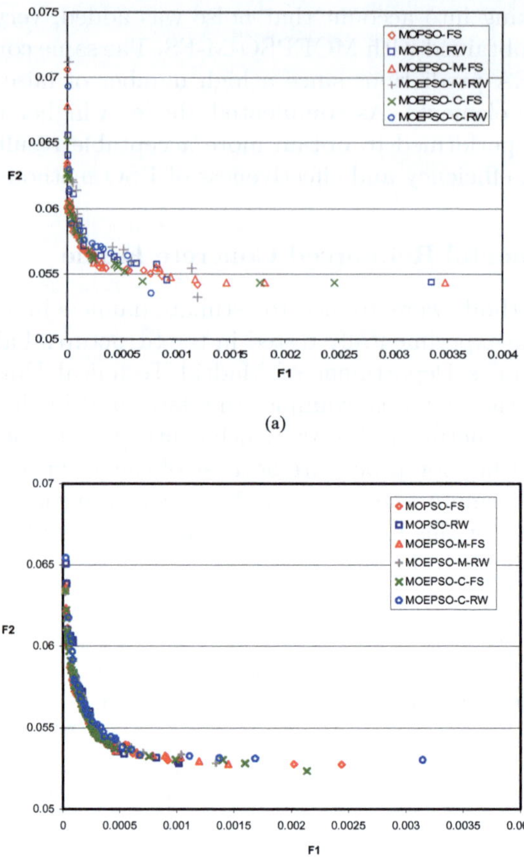

(a)

(b)

Fig. 8.9. Pareto fronts for the six versions of the PSO algorithm: Frame. (a) 30 runs; (b) 100 runs

Firstly, the six versions of PSO algorithms were used for solving the multiobjective problem. The same parameters and the same number of runs, 30 and 100, as in the previous example were used except the number of iterations per run which was increased to 150 given the higher difficulty of the problem. The nodes of the finite element mesh were made coincident with the location of the accelerometers resulting finally 12 elements, 2 at each column and 8 at the beam. Figs. 8.9, 8.10 and 8.11 show the Pareto front. Tables 8.6,8.7,8.8 and 8.9 show the average density values and the coverage. In this problem, more complex than the previous one, the average performance of MOEPSO-M-FS is the best among the six algorithms adopted when 30 runs were performed although its superiority is not so evident. However, as in the

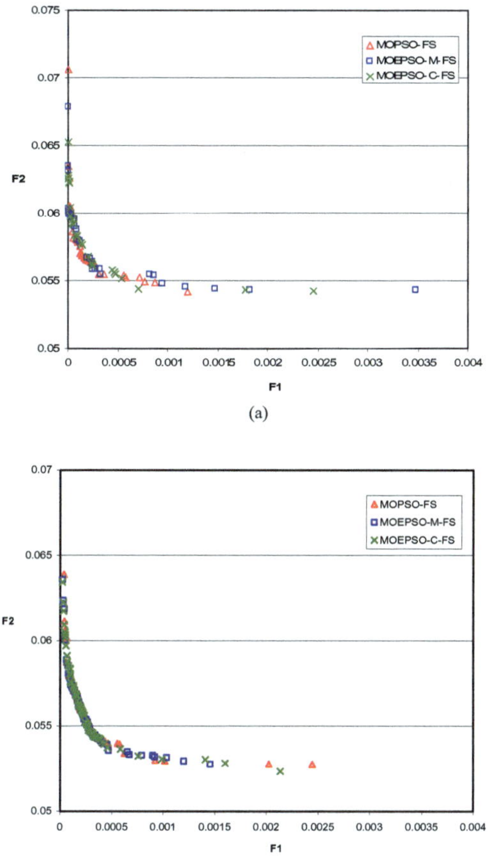

(a)

(b)

Fig. 8.10. Pareto fronts for the three versions of the PSO algorithm based on FS: Frame. (a) 30 runs; (b) 100 runs

Table 8.6. Average density values and Pareto front sizes: Frame (30 runs)

	MOPSO-FS	MOPSO-RW	MOEPSO-M-FS	MOEPSO-M-RW	MOEPSO-C-FS	MOEPSO-C-RW
Average Density Value	13	6.55	7.65	2.81	7.45	7.75
Pareto front size	48	16	36	12	29	15

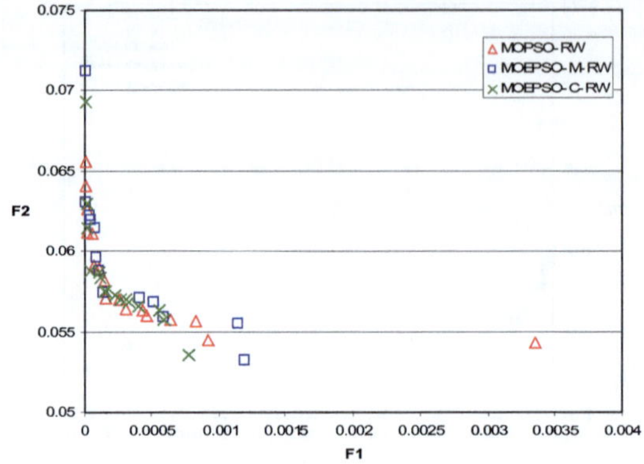

(a)

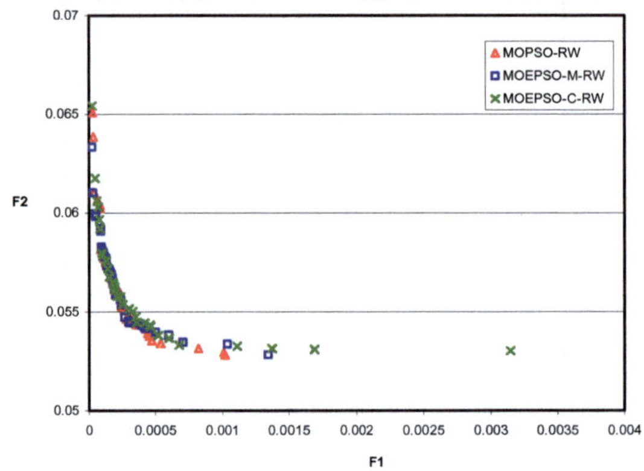

(b)

Fig. 8.11. Pareto fronts for the three versions of the PSO algorithm based on RW: Frame. (a) 30 runs; (b) 100 runs

Table 8.7. Average density values and Pareto front sizes: Frame (100 runs)

	MOPSO-FS	MOPSO-RW	MOEPSO-M-FS	MOEPSO-M-RW	MOEPSO-C-FS	MOEPSO-C-RW
Average Density Value	18.33	8.95	23.38	9.05	15.14	9
Pareto front size	72	39	83	39	72	35

Table 8.8. C metrics to measure the coverage of two sets of solutions: Frame. X′: First column; X″: First line (30 runs)

	MOPSO-FS	MOPSO-RW	MOEPSO-M-FS	MOEPSO-M-RW	MOEPSO-C-FS	MOEPSO-C-RW
MOPSO-FS	N/A	0.94	0.38	0.92	0.5	0.94
MOPSO-RW	0	N/A	0.08	0.61	0	0.56
MOEPSO-M-FS	0.28	0.82	N/A	0.84	0.13	0.75
MOEPSO-M-RW	0.04	0.12	0.08	N/A	0.1	0.12
MOEPSO-C-FS	0.3	0.82	0.35	0.85	N/A	0.75
MOEPSO-C-RW	0.06	0.41	0.27	0.69	0.2	N/A

Table 8.9. C metrics to measure the coverage of two sets of solutions: Frame. X′: First column; X″: First line (100 runs)

	MOPSO-FS	MOPSO-RW	MOEPSO-M-FS	MOEPSO-M-RW	MOEPSO-C-FS	MOEPSO-C-RW
MOPSO-FS	N/A	0.03	0.01	0.03	0	0
MOPSO-RW	0	N/A	0.01	0.03	0	0
MOEPSO-M-FS	0	0.03	N/A	0.03	0	0
MOEPSO-M-RW	0	0.03	0.01	N/A	0	0
MOEPSO-C-FS	0	0.03	0.01	0.03	N/A	0
MOEPSO-C-RW	0	0.03	0.01	0.03	0	N/A

previous example, for 30 runs, it clear that FS selection scheme performs better than RW selection scheme. When 100 runs are performed the FS selection scheme is superior regarding the average density value and the Pareto front size. However, none scheme dominates regarding the coverage since all the Pareto fronts practically overlap.

A comparison has been also performed in this example between one PSO based algorithm, MOEPSO-M-FS, and one EA, SPGA. Fig. 8.12 shows the Pareto fronts and again, as in the previous example, the advantage of PSO algorithm over EA is evident. Damage predictions in the beam part of the frame with both algorithms are also shows in Fig.8.13. MOEPSO-M-FS identifies perfectly the damage regions at the ends of the beam and at the midspan of the beam. However, SPEA estimations are enough erroneous which demonstrates again that this algorithm would need a much higher number of generations to reach a good prediction.

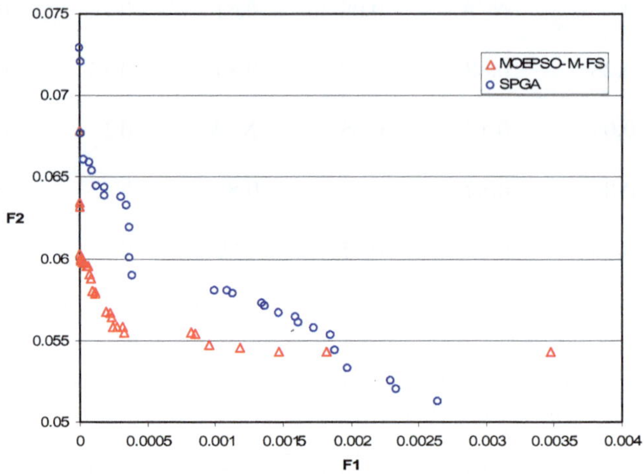

Fig. 8.12. Comparison of Pareto fronts between SPGA and MOEPSO-M-FS

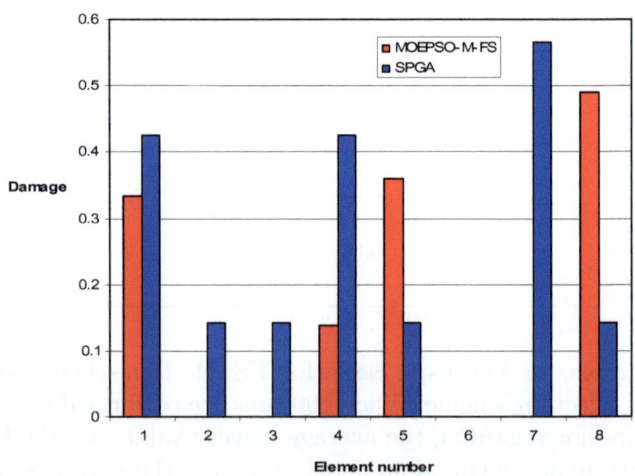

Fig. 8.13. Comparison of damage predictions among MOEPSO-M-FS and SPGA for the beam part of the frame

8.7 Conclusions

Different PSO-based algorithms have been proposed and tested in multiobjective damage identification problems. The numerical results demonstrate the robustness and feasibility of the proposed methods for damage identification of different structures. To the knowledge of the authors, it is the first time that these optimization schemes have been applied for solving this kind of problem, very important from an economic and safety point of view. Moreover, even though very few works have been reported to date, results demonstrate that PSO-based algorithms can become highly competitive with other algorithms in solving multiobjective problems, such as EAs. In our particular problem, their high convergence speed and feasibility of implementation make them an ideal candidate to implement a structural health monitoring technique in real time.

Acknowledgements

This work is supported by the Ministry of Education and Science of Spain through project BIA2007-67790.

References

1. Ewins, D.J.: Modal testing: Theory and practice. Wiley, New York (1984)
2. Doebling, S.W., Farrar, C.R., Prime, M.B.: A summary review of vibration-based damage identification methods. Shock Vibration 30(2), 91–105 (1998)
3. Yan, Y.J., Cheng, L., Wu, Z.Y., Yam, L.H.: Development in vibration-based structural damage detection technique. Mechanical Systems and Signal Processing 21, 2198–2211 (2007)
4. Brownjohn, J.M.W., Xia, P.Q., Hao, H., Xia, Y.: Civil structure condition assessment by FE model updating methodology and case studies. Finite Elements in Analysis and Design 37, 761–775 (2001)
5. Perera, R., Torres, R.: Structural damage detection via modal data with genetic algorithms. Journal of Structural Engineering ASCE 132(9), 1491–1501 (2006)
6. Friswell, M.J.: Damage identification using inverse methods. Philosophical Transactions of the Royal Society 365(1851), 393–410 (2007)
7. Haralampidis, Y., Papadimitriou, C., Pavlidou, M.: Multiobjective framework for structural model identification. Earthquake Engineering and Structural Dynamics 34, 665–685 (2005)
8. Perera, R., Ruiz, A., Manzano, C.: An evolutionary multiobjective framework for structural damage localization and quantification. Engineering Structures 29(10), 2540–2550 (2007)
9. Fonseca, C.M., Fleming, P.J.: An overview of evolutionary algorithms in multiobjective optimization. Evolutionary Computation 3, 1–16 (1995)
10. Coello, C.A., Van Veldhuizen, D.A., Lamont, G.B.: Evolutionary algorithms for solving multiobjective problems. Kluwer Academic Publishers, New York (2002)

11. Lagaros, N.D., Plevris, V., Papadrakakis, M.: Multi-objective design optimization using cascade evolutionary computations. Computer Methods for Applied Mechanics and Engineering 194, 3496–3515 (2005)
12. Coello, C.A.: Recent trends in evolutionary multiobjective optimization. In: Abraham, A., Jain, L.C., Goldberg, R. (eds.) Evolutionary Multiobjective Optimization:Theoretical Advances and Applications. Springer, London (2005)
13. Kennedy, J., Eberhart, R.C.: Particle swarm optimization. In: Proceedings of IEEE International Conference on Neural Networks, Piscataway, New Jersey, pp. 1942–1948 (1995)
14. Zhang, C., Shao, H., Li, Y.: Particle swarm optimization for evolving artificial neural network. In: Proceedings of the IEEE International Conference on Systems, Man and Cybernetics, pp. 2487–2490 (2000)
15. Kennedy, J., Eberthart, R.C., Shi, Y.: Swarm Intelligence. Morgan Haufman Publishers, San Francisco (2002)
16. Eberthart, R.C., Shi, Y.: Particle swarm optimization: developments, applications and resources. In: Proceedings of the 2001 Congress on Evolutionary Computation, Seoul, pp. 81–86 (2001)
17. Abido, M.A.: Optimal design of power system stabilizers using particle swarm optimization. IEEE Transactions on Energy Conversion 17(3), 406–413 (2002)
18. Agrafiotis, D.K., Cedeno, W.: selection for structure-activity correlation using binary particle swarms. Journal of Medicinal Chemistry 45(5), 1098–1107 (2002)
19. Coello, C.A., Lechuga, M.S.: MOPSO: A proposal for multiple objective particle swarm optimization. In: Proceedings of the IEEE Congress on Evolutionary Computation, Honolulu, Hawaii, pp. 1677–1681 (2002)
20. Hu, X., Eberhart, R.: Multiobjective optimization using dynamic neighborhood particle swarm optimization. In: Proceedings of the IEEE Congress on Evolutionary Computation, Honolulu, Hawaii, pp. 1677–1681 (2002)
21. Parsopoulos, K.E., Vrahatis, M.N.: Particle swarm optmization method in multiobjective problems. In: Proceedings of the 2002 ACM Symposium on Applied Computing, Madrid, pp. 603–607 (2002)
22. Hu, X., Shi, Y., Eberhart, R.: Recent advances in particle swarm. In: IEEE Congress on Evolutionary Computation, Portland, Oregon, pp. 90–97 (2004)
23. Coello, C.A., Pulido, G.T., Lechuga, M.S.: Handling multiple objectives with particle swarm optimization. IEEE Transactions on Evolutionary Computations 8(3), 256–279 (2004)
24. Srinivasan, D., Seow, T.H.: Particle swarm inspired evolutionary algorithm (PS-EA) for multi-criteria optimization problems. In: Abraham, A., Jain, L.C., Goldberg, R. (eds.) Evolutionary Multiobjective Optimization:Theoretical Advances and Applications. Springer, London (2005)
25. Au, F.T.K., Cheng, Y.S., Tham, L.G., Bai, Z.Z.: Structural damage detection based on a micro-genetic algorithm using incomplete and noisy modal test data. Journal of Sound and Vibration 259(5), 1081–1094 (2003)
26. Friswell, M.I., Penny, J.E.T., Garvey, S.D.: A combined genetic and eigensensitivity algorithm for the location of damage in structures. Computers and Structures 69, 547–556 (1998)
27. Jaishi, B., Ren, W.X.: Damage detection by finite element model updating using modal flexibility residual. Journal of Sound and Vibration 290, 369–387 (2006)

28. Ho, S.L., Yang, S., Ni, G., Lo, E.W.C., Wong, H.C.: A particle swarm optimization-based method for multiobjective design optimizations. IEEE Transactions on Magnetics 41(5), 1756–1759 (2005)
29. Morse, J.N.: Reducing the size of the nondominated set: Pruning by clustering. Computational Operations Research 7(1-2), 55–66 (1980)
30. Goldberg, D., Richardson, J.J.: Genetic algorithms with sharing for multimodal function optimization. In: Proceedings of the Second International Conference on Genetic Algorithms on Genetic algorithms and their application, Cambridge, Massachusetts, pp. 41–49 (1987)
31. Zitzler, E., Thiele, L.: Multiobjective evolutionary algorithms: A comparative case study and the strength Pareto approach. IEEE Transactions on Evolutionary Computation 3(4), 257–271 (1999)
32. Agrawal, S., Dashora, Y., Tiwari, M.K., Son, Y.J.: Interactive particle swarm: A Pareto-adaptive metaheuristic to multiobjective optimization. IEEE Transactions on Systems, Man and Cybernetics C Part A: Systems and Humans 38(2), 258–271 (2008)
33. Horn, J., Nafpliotis, N., Goldberg, D.E.: A niched Pareto genetic algorithm for multiobjective optimization. In: Proceedings of the 1st IEEE Conference on Computation Evolutionary, vol. 1, pp. 82–87 (1994)

Author Index

Index

MIX
Papier aus verantwortungsvollen Quellen
Paper from responsible sources
FSC® C105338

If you have any concerns about our products,
you can contact us on
ProductSafety@springernature.com

In case Publisher is established outside the EU,
the EU authorized representative is:
Springer Nature Customer Service Center GmbH
Europaplatz 3, 69115 Heidelberg, Germany

Printed by Libri Plureos GmbH
in Hamburg, Germany